The Princess & THE PATRIOT

The Princess & THE PATRIOT

Ekaterina Dashkova,
Benjamin Franklin, and the
Age of Enlightenment

Sue Ann Prince, Editor

AMERICAN PHILOSOPHICAL SOCIETY

Philadelphia, 2006

Transactions of the American Philosophical Society
Held at Philadelphia
For Promoting Useful Knowledge
Volume 96, Part 1

Published for the exhibition *The Princess & the Patriot:
Ekaterina Dashkova, Benjamin Franklin, and the Age of Enlightenment*

Presented at the American Philosophical Society
Philosophical Hall, Philadelphia, Pennsylvania
February 2006 through December 2006

Library of Congress Cataloging-in-Publication Data

The princess & the patriot : Ekaterina Dashkova, Benjamin Franklin , and the Age of
Enlightment / Sue Ann Prince, editor.
    p. cm. — (Transactions of the American Philosophical Society, held at Philadelphia for
promoting useful knowledge ; vol. 96, pt. 1)
   The fourth exhibition organized by the Museum of the American Philosophical Society in
celebration of the 300th anniversary of the birth of its founder, Benjamin Franklin; from
Feb. to Dec. 2006.
   Includes bibliographical references and index.
   ISBN-13: 978-0-87169-961-9 (pbk.)
   ISBN-10: 0-87169-961-3 (pbk.)
   1. Dashkova. E. R. (Ekaterina Romanovna), knëìginëì, 1743–1810 – Exhibitions. 2.
American Philosophical Society–Biography–Exhibitions. 3. Franklin, Benjamin,
1706–1790 – Exhibitions. 4. Princesses–Russia–Biography–Exhibitions. 5.
Intellectuals–Russia–Biography–Exhibitions. 6. Russia–Court and
courtiers–Biography–Exhibitions. I. Title; Princess and the patriot. II. Prince, Sue Ann.
III. American Philosophical Society. Museum. IV. Transactions of the American
Philosophical Society ; vol. 96, pt. 1.

DK169.D3P75 2005
947'.063092–DC2   2005057219

Designed by Ann Antoshak   US ISSN: 0065-9738

FRONT COVER and FRONTISPIECE   **Left:** Alexei Nesterovich Maximow, *Portrait of Ekaterina R. Dashkova as Director of the Imperial Academy of Sciences and Arts*, 2003. Replica of an eighteenth-century portrait by an unknown artist; the original is in the collection of the State Historical Museum, Moscow. Oil on canvas. **Right:** Charles Willson Peale, *Portrait of Benjamin Franklin*, 1772. Copy of a 1767 painting by David Martin. Oil on canvas. **Both:** American Philosophical Society. Photographs by Frank Margeson.

Table of Contents

Foreword

The Princess and the Patriot: Ekaterina Dashkova, Benjamin Franklin, and the Age of Enlightenment is the title not only of this volume but also of an exhibition held at the Museum of the American Philosophical Society in Philosophical Hall, Philadelphia, from February through December 2006. The fourth exhibition organized by the Museum since the Society's new outreach program was established in 2001, it is a major contribution to the Society's celebration of the 300th anniversary of the birth of its founder, Benjamin Franklin.

It is also an occasion to celebrate Ekaterina Romanovna Dashkova, an eighteenth-century woman of science who became the Society's first female member, elected in 1789.

She was the Director of the Imperial Academy of Sciences and Arts under Catherine the Great. It is the first exhibition on Dashkova in the United States and the first to juxtapose the manifestation of the Enlightenment in North America to that of Catherinian Russia. Included in the show are portraits, jewelry, porcelain, engravings, maps, memoirs, letters, and scientific objects. Many of the objects are being borrowed from other institutions including the Archives of the Russian Academy of Sciences, the Peter the Great Museum of Anthropology and Ethnography (Kunstkamera) in St. Petersburg, the State Historical Museum in Moscow, the State Hermitage Museum in St. Petersburg, and the Royal Irish Academy in Dublin.

The first section of the exhibition explores the lives of these two luminaries in their own very different worlds—Dashkova's St. Petersburg, a grand imperial city on the Neva River, and Franklin's Philadelphia, a bustling new port on the Delaware River, the birthplace of a new nation. The displays then consider the context that Dashkova and Franklin encountered in their travels and sojourns in Western Europe, leading up to their one and only face-to-face encounter, which took place in Paris in 1781. The remainder of the exhibition is organized according to three major Enlightenment ideals that were espoused by Dashkova and Franklin: the pursuit of knowledge through the use of reason, the right to liberty and equality, and the attainment of virtue through self-improvement.

This book, which will remain as a record of the exhibition long after it closes at the end of 2006, includes eight essays and a checklist. Neither it nor the exhibition would be possible without the generous support of many foundations and individuals who are helping us realize our mission of interpreting to a broad audience the intersections of history, art, and science. We are especially indebted to the William Penn Foundation, the Institute of Museum and Library Services, the Pennsylvania Historical and Museum Commission, the Heritage Philadelphia Program funded by the Pew Charitable Trusts, the Pennsylvania Humanities Council, the Richard S. Lounsbery Foundation, the Patchwork Charitable Foundation in memory of R. Stewart Rauch and Frances B. Rauch, the Trust for Mutual Understanding, the Arcadia Foundation, and the Florence R. C. Murray Charitable Trust. We are extremely grateful for their interest in our new public outreach program, and for their willingness to support it.

Baruch S. Blumberg, President
American Philosophical Society

Acknowledgments

The Princess and the Patriot: Ekaterina Dashkova, Benjamin Franklin, and the Age of Enlightenment is the most ambitious exhibition the Society has organized in Philosophical Hall since it opened its doors to the public in 2001. It is our first show to entail loans from foreign institutions and our second to include a companion book. I owe an enormous debt to the many lenders, scholars, curators, donors, and staff members who have played a role in realizing this ambitious project.

First I would like to pay tribute to Frank H. T. Rhodes, former American Philosophical Society president, and Alexander G. Bearn, former executive officer, for supporting the idea for the exhibition project in 2002, long before it was fully conceptualized. They took a risk in heartily endorsing the celebration of an eighteenth-century woman of science, along with Franklin, for the Benjamin Franklin Tercentenary exhibition in Philosophical Hall. Baruch S. Blumberg, who became president of the Society in 2005, has been equally enthusiastic, and co-executive officers Mary Maples Dunn and Richard S. Dunn have also encouraged the realization of the project despite the many financial challenges it has engendered. Members of the Museum advisory committee—Henry A. Millon (Society Curator and Committee Chair), along with David Brigham, Jonathan Brown, Elizabeth Cropper, Sheldon Hackney, Sue Johnson, and Charles A. Ryskamp—have shared their wisdom and provided crucial practical support whenever necessary. I am also indebted to other APS members such as Loren R. Graham, who has helped in numerous ways, from negotiating in Moscow for the commissioning of a copy of an eighteenth-century Dashkova portrait to sharing his knowledge of the history of Russian science.

Special thanks and admiration go to all the scholars who contributed essays to this volume, most especially those whose work required new and original research in Russian archival repositories. In addition to the authors, there were many other professionals outside the APS who played significant roles at various stages of the project. Greg Guroff, Marilyn P. Swezey, and Ann Van Devanter Townsend all offered valuable assistance during the initial planning phases; and interpretive consultant Janet Kamien provided invaluable insights as we were conceptualizing the exhibition, as did Alexander Woronzoff-Dashkoff, a descendant of Dashkova, and his wife Catherine. Alexander Shedrinsky facilitated my work in Russia in numerous ways over the past three years—opening doors into collections, making contact with appropriate staff members at numerous institutions, translating documents, and, in general, ensuring that all travel and negotiations were accomplished as quickly and smoothly as possible.

To the private and institutional lenders, many of whom have shown unusual generosity in helping realize this exhibition, I am also enormously grateful. I especially salute those colleagues abroad who generously allowed me access to their collections and gave of their time to work out all the details of our loan requests. Included are Alexander Ivanovich Shkurko and Tamara Grigoriyevna Igumnova of the State Historical Museum in Moscow; Irina Vladimirovna Tunkina and Irina Mikhailovna Shchedrova at the Archives of the Russian Academy of Sciences in St. Petersburg; Yuri K. Christov, Julia A. Kupina, and Tatyana M. Moisseeva at Peter the Great Museum of Anthropology and Ethnography (Kunstkamera) of the Russian Academy of Sciences; Mikhail Borissovich Piotrovsky, Vladimir Yurievich Matveev, and Nadezhda Bronislavovna Petrussevich at the State Hermitage Museum; and Siobhán Fitzpatrick and her staff at the Royal Irish Academy Library in Dublin.

In the United States, we also worked closely with many scholars, curators, and registrars at lending institutions. Staff members at Hillwood Museum & Gardens, including Frederick J. Fisher, David Johnson, Karen Kettering, Anne Odom, and Ruthann Uithol, have been enormously generous throughout the project not only in lending objects from their museum's incomparable collections but also in sharing their knowledge, expertise, and moral support. At the New York Public Library, Edward Kasinec's enthusiasm for the project, along with that of HeeGwone Yoo and Roseann Panebianco, was most encouraging. I also want to thank Sara W. Duke of the Library of Congress, who helped us find many splendid prints there, and her colleagues John R. Hébert, James Hutson, Mark Dimunation, and Rachel Waldron. As for our two previous exhibitions, our friends and colleagues here in Philadelphia agreed to lend some of their institutions' finest treasures in order to enrich our current show. Our warmest thanks to D. James Baker, Lucinda McDade, The Academy of Natural Sciences; Elliot L. Shelkrot, Richard C. Boardman, Free Library of Philadelphia; Dennis M. Wint and John Alviti, The Franklin Institute; David Moltke-Hansen, Lee Arnold, and Max Moeller, The Historical Society of Pennsylvania; John C. Van Horne, James N. Green, Sarah J. Weatherwax, and Jennifer Ambrose, The Library Company of Philadelphia; J. Justin and Gwen Ragsdale, "Lest We Forget" A Private Collection of Slavery Artifacts and Jim Crow Memorabilia; Janet Evans, The Pennsylvania Horticultural Society; Anne d'Harnoncourt, Kathryn B. Hiesinger, and Donna Corbin, Philadelphia Museum of Art; and Derick Dreher, Judith Guston, Karen Schoenewaldt, and Elizabeth E. Fuller, The Rosenbach Museum & Library. Other colleagues outside of Philadelphia who also gave of their time and expertise include Ellen Kuhfeld, The Bakken Library and Museum, Minneapolis; Roberta Zhongi, Rare Books and Manuscripts Department, The Boston Public Library; Saundra Taylor, The Lilly Library, Indiana University, Bloomington; Tamara C. Barnes, Salem County Historical Society, Salem, N. J.; Peggy Asbury, Alaska and Polar Regions Collections, Elmer E. Rasmuson Library, The University of Alaska Fairbanks; Rebecca Dunkle, University Library, University of Michigan, Ann Arbor; William R. Johnston, Liz Flood, and Betsy Dahl, The Walters Art Museum, Baltimore; and one anonymous lender.

Three independent colleagues who work with us on a freelance basis deserve special mention. I have worked with them not only on this exhibition but for all or most of my years at the Society. Exhibition designer Stephen Saitas once again combined his infallible sense of space and eye for detail with an indomitable ability to choreograph the many people involved in fabricating exhibit mounts and installing objects, ultimately turning our ideas into absorbing visual displays. Ann Antoshak, our graphic designer, provided striking graphic design for every aspect of the project. Her free-flowing creativity resulted in a host of intriguing design solutions for everything from logo and letterhead to signage and interpretive panels. Working with both these designers to implement our vision was truly a pleasure. Elaine Wilner, our public relations professional, cheerfully and generously shared her prolific ideas and insights, and used her prodigious savvy to publicize the exhibition, continuing her dedication to getting the word out about the Museum and the Society through the astute placement of articles and media spots.

Among the many APS staff members outside the Museum program who have been involved, I am grateful to librarian Martin L. Levitt and his team—Robert S. Cox (now at the University of Massachusetts Amherst), Charles B. Greifenstein, Roy E. Goodman, Richard Shrake, and Valerie-Anne Lutz especially, who showed unfailing good cheer as the Museum staff perused innumerable documents and books in search of materials for the show. Val's unusual ability to locate wayward documents and remain energized and cheerful throughout the tracking of endless registrarial details was invaluable. Conservators Anne E. Downey and Denise Carbone ably and skillfully assessed, cleaned, conserved, and prepared paper objects for the exhibition, allowing the Society literally to put its best face forward, while free-lance framer and preparator Bill Severson tirelessly and generously gave of his time to prepare, frame, and in some cases install objects for display.

Other staff members who have played crucial roles include photographer Frank Margeson, whose work not only helped us as we conceptualized the exhibition, but also as we shared it with the public. Carl F. Miller and John Wolfe have offered invaluable assistance in all things pertaining to the budgeting process, while Nanette Holben and Celeste DiNucci worked tirelessly to secure the funds that made the exhibition possible. Thanks also goes to Frank Sabatini and his maintenance staff for their support during the crucial transition period between exhibitions. Mary C. McDonald, along with Celeste Bivings, shepherded this volume through the complex publication process, and free-lance editor Nicole Amoroso made perceptive suggestions and changes that improved every page.

To my own superb Museum staff there are no words that can fully express my gratitude not only for their excellent work but also for their loyalty and their willingness to go the extra mile—to push just a little harder to get things right, whether it means rewriting a label or working as a team to finish a job. A prodigious amount of work was necessary for planning and implementing this show, and they never ceased surprising me with their creative ideas and ability to make things happen. As this project has been in the making for over four years, full- and part-time staff members whom I wish to acknowledge personally include former and current curatorial assistants Mary Catherine (Katie) Wood and Rosemary K. McCarthy respectively, who skillfully assisted in every aspect of planning and implementation; my superb research associates Jane E. Boyd and Erin McLeary, both of whom have been an enthusiastic and integral part of the exhibition team for most of the Museum's history; Enlightenment scholar and researcher Jenny Furlong; native Russian speaker Masha Shardakova, who early on in the development of the exhibition researched Russian documents and provided translation where needed; and finally, our curator of Museum education Mary Teeling, whose superb sense of how to turn the curatorial research work into meaningful programming for our visitors provides the essential link between the scholarly research we do and the audience experience. Thanks also goes to our contract registrarial staff, including Mary Grace Wahl, who stepped into the project in midstream and skillfully ushered it through several critical junctures, and Stephanie Wiener, first a registrarial intern and then assistant registrar, who learned on her feet and skillfully tracked all the objects and implemented the shipping details. Finally, I want to express my great appreciation to Irina Dubinina, our translator and advisor on all things Russian over the past two years; we drew on her superb language skills for crucial translations of publications, contracts, and letters, late-night phone calls to Russia, and arrangements for hosting our Russian visitors, all of which she accomplished with unusual grace and tact.

For these and all other colleagues and supporters too numerous to mention, I offer my heartfelt thanks for their assistance in solving problems, offering moral support, and generously giving of their knowledge and skills. They all remind me of how the success of such a complex project as this one is always the result of collaboration—the untold efforts and ideas of many. I am ever so grateful!

Introduction
A Meeting of Minds

On February 3, 1781, Benjamin Franklin met Russian Princess Ekaterina Romanovna Dashkova at the Hôtel de la Chine in Paris. It was an unlikely meeting between two extraordinary Enlightenment figures from opposite sides of the world. She was a forthright thirty-seven-year-old noblewoman from a powerful Russian monarchy; he was an elderly statesman from a new democracy in the making. Though they would never meet again, they honored each other in a unique way as a result of their encounter. In 1789, Franklin nominated Dashkova, who by then was director of the Imperial Academy of Sciences and Arts in St. Petersburg, to the American Philosophical Society in Philadelphia. She became its first female member.

He, in turn, as a natural philosopher—the eighteenth century's word for scientist—and founder of what was then America's only learned society, would become the first American member of the Russian Academy.

Dashkova and Franklin were exemplars of Enlightenment thought as well as unusually colorful personalities whose accomplishments and unexpected behaviors turned them into legends in their respective countries. Dashkova, who broke all stereotypes about princesses, was a sometimes-outrageous lady-in-waiting who spurned court attire for men's clothing, participated in a coup d'état that overthrew a czar, and became the first woman in the world to lead a learned society. Franklin, no ordinary citizen-patriot, was a self-made entrepreneur at home and a savvy diplomat abroad. He, like Dashkova, had the verve to defy court fashion, wearing a fur cap "among the Powder'd Heads of Paris."[1] He also had the foresight to found a scholarly society in a young colonial city that by European standards was little more than a cultural backwater.

In conjunction with the tercentenary celebration of Franklin's birth in 2006, the American Philosophical Society has organized an exhibition titled *The Princess and the Patriot: Ekaterina Dashkova, Benjamin Franklin, and the Age of Enlightenment*. It will be the first exhibition in the United States to explore the life and work of Dashkova, and the first to juxtapose her with Franklin. This book, the 489th volume of the Society's *Transactions*, which have been published continuously since 1771, is being issued as a companion to the exhibition.

Dashkova, Franklin, and the Challenges of the Enlightenment

Dashkova met and befriended the future Catherine the Great when she was only fifteen, and at sixteen she became a princess when she married Prince Mikhail I. Dashkov.[2] After Dashkova's involvement in the 1762 coup that overthrew Catherine's husband, Peter III, she served as lady-in-waiting to the new empress, and twenty years later, in 1783, she was appointed director of Russia's scientific academy. That same year she became the founding president of the Imperial Academy of the Russian Language, where she published the first Russian dictionary.

Dashkova's love for learning was sparked at an early age. She amassed a library of some nine hundred books as a teenager, read Montesquieu's *Esprit des lois* and other seminal Enlightenment texts, and learned five languages. Later, she would make two extensive trips through Europe, meeting with such Enlightenment *philosophes* as Voltaire and Denis Diderot. Like Franklin, she explored liberal political concepts and struggled with the challenges that the new Enlightenment ideals were presenting to individuals and governments.

In this volume, as in the exhibition, the lives of Dashkova and Franklin offer a lens through which to view the tensions and contradictions that arose when new Enlightenment ideals encountered practices that were deeply entrenched in the political, economic, and social realities of the time. Dashkova's and Franklin's belief in the importance of engendering and promoting useful knowledge for the betterment of humankind is developed in a section of the show titled *"The Pursuit of Knowledge and the Use of Reason."* Included are letters and other documents that exemplify the extensive international exchange of scientific ideas that occurred during the Enlightenment. Also exhibited are artifacts relating to scientific experiments and expeditions sponsored by the Russian Academy of Sciences and the American Philosophical Society. In the first essay of this book, "Arduous and Delicate Task: Princess Dashkova, the Academy of Sciences, and the Taming of Natural Philosophy," Michael Gordin ably and eloquently explores how Dashkova's tenure at the Russian Academy effected a crucial transformation in the way natural inquiry was conducted. He argues that her style of leadership separated the pursuit of knowledge from institutional administration, thereby changing the relationship between the academicians and their sovereign. Further noting that Dashkova "stood at the very dawn of the metamorphosis of 'natural philosophy' into 'science,'" he shows that although her work exemplified Enlightenment principles in many ways, she paradoxically participated in a crucial transition that ultimately transformed the eighteenth century's notion of natural philosophy into the nineteenth century's more professionalized and specialized scientific practice.

"*Liberty and Equality*," also an Enlightenment ideal, is the subject of another section of the exhibition. It documents Dashkova's and Franklin's struggle to reconcile their evolving ideas about universal human rights—liberty and freedom for all—with the practices of serfdom and slavery in their respective countries as well as in their own lives. Included are engravings and watercolors documenting practices of human bondage along with abolitionist images and texts that reveal the growing outrage against tyranny in all forms. In an essay titled "Liberty Postponed: Princess Dashkova and the Defense of Serfdom," Michelle Marrese thoroughly and sensitively explores the delicate position of the Russian nobility in relation to serfdom. She argues that while Dashkova's defense of the practice seems to modern scholars "at odds with her promotion of the Enlightenment in Russia," her position was based largely on the context in which the Russian nobility operated. She was part of a noble class whose own liberty, property rights, and social status in relation to the sovereign were without guarantee; in other words, the nobility itself did not have political freedom. Thus, according to Marrese, "despite her genuine interest in ameliorating the lot of the Russian peasantry, Dashkova's primary goal was to shore up the security and privileges of the ruling class." Furthermore, Dashkova believed that as long as there was a Russian sovereign who could exercise arbitrary power, noble proprietors were essential because their serfs needed protection from corrupt officialdom. She concludes: "Dashkova's failure to advocate the abolition of serfdom was emblematic of the tension in Enlightenment philosophy between the celebration of freedom and the reality of governing an imperfect society."

The final Enlightenment theme explored in the exhibition is "*Virtue and Self-*

Improvement." Exhibited are the autobiographical memoirs that Dashkova and Franklin wrote respectively, along with other materials that explore the eighteenth-century notion of virtue; a miniature chess set, for example, is juxtaposed with Franklin's article "On the Morals of Chess," and manuscripts penned by Dashkova on the subject of personal and civic virtue are also on display. In addition to revealing notions of living an ethical life, this section also uncovers how both Dashkova and Franklin used their autobiographical writings to fashion virtuous personae for public consumption, even if they did not always attain such virtue in practice. In Marcus Levitt's perceptive essay, titled "Virtue Must Advertise: Self-Presentation in Princess Dashkova's Memoirs," the author describes Dashkova's memoirs "as an attempt to rescue her public image from oblivion or worse, misrepresentation." Levitt deftly argues that Dashkova's stance in her memoirs resembles that of a heroine in a classical tragedy, one "surrounded by a world that cannot possibly equal or appreciate her." The contradiction in *Mon histoire* between Dashkova's strongly authoritative sense of *self* and a seemingly contradictory assertion of *selflessness* also, according to Levitt, recalls the classicist model of selfhood, echoing the moral discourse of the Russian tragic stage and enacting its own analogous display of virtue.

These first three essays focus largely on Dashkova, who is little known to an American audience. The fourth work, however, does not correspond to any one section of the exhibition, but rather explores major themes in Franklin's life and the relationship of his work to that of Dashkova and the Enlightenment as a whole. In this compelling essay, written by Karen Duval and titled "A Man Made to Measure: Benjamin Franklin, American *Philosophe*," the author discusses Franklin as "the *philosophe* from America" who, "in the eyes of the French and much of Europe . . . had gone further and done something no other *philosophe* had dared to do, namely, to put theory into practice and declare independence."

Duval's essay is followed by four shorter documentary essays that draw on archival sources to address specific aspects of Dashkova's life and work: the library she amassed as a teenager, her personal relationships with others, her love for England, and the many faces she presents in her portraits. Written by Alexander Woronzoff-Dashkoff, Svetlana Dolgova, Anthony Cross, and Elena Stolbova respectively, these essays are revised or expanded versions of previously published works that appeared in Russian-language publications or in academic journals. The purpose of their inclusion here is to make them more readily available to a broad English-language readership.

In the process of comparing the contributions of two larger-than-life luminaries of the eighteenth century, the exhibition and book offer insight into how the Enlightenment manifested itself in two outposts on either side of Europe's intellectual centers. The juxtaposition of Dashkova and Franklin and of Russia and America allows each person and each context to serve as a foil for the other. Some surprising similarities and differences emerge in the areas of science, politics, and culture.

Dashkova's complex and multiple roles—as friend of Catherine II, academy director, widowed mother of two, and proprietess of large estates—highlight the absence of women leaders in colonial America and the early Republic. Her stance on serfdom reveals a striking overall resemblance to justifications used to explain away slavery in the New World despite differences between the two forms of servitude. The abuses of power and the many court intrigues that Dashkova encountered in an autocratic monarchy provide a basis for understanding Franklin's evolution from an ardent royalist to a revolutionary statesman who helped free his native land from old-world tyranny.

Despite their unique struggles and contexts, however, both shared a belief in progress through natural philosophy, both found it difficult to reconcile the entrenched practices of human bondage with the emerging eighteenth-century ideals of liberty and equality for all people, and both enacted a notion of personal virtue and civic responsibility for the betterment of human society. These commonalities in their worldviews reveal how Enlightenment ideals spread from one end of the western world to the other, transcending differences of age, gender, country of origin, and political context.

This project, which includes an exhibition, public programming, high-school partnerships, and this book, not only explores how these two exemplary figures struggled to put their ideals into practice—with varying degrees of success—but also highlights how the same ideals are as challenging today as they were in the eighteenth century. Science is not always used for benevolent purposes; slavery in a variety of forms still exists throughout the world; and notions of personal ethics and civic virtue are often abandoned in favor of personal gain. These yet-unrealized Enlightenment ideals are what make this exhibition and book not only relevant to historians of the eighteenth century but to all of us today.

Sue Ann Prince
Director of Exhibitions and Collections Museum
Curator of *The Princess and the Patriot*

Notes

1. Benjamin Franklin to Polly Stevenson, September 14, 1767, January 12, 1777 (Benjamin Franklin, *The Papers of Benjamin Franklin*, eds. Leonard W. Labaree et al. [New Haven, 1959–], vol. 14, pp. 254–55; vol. 23, pp. 155–56).

2. The Russian words for "prince" and "princess" originally referred to the ruler of a principality and his wife (before principalities were united into the one state that came to be known as Russia). Subsequently, "prince" came to be used for nobles who traced their lineage to the ruling families of the principalities. Since this title and that of princess were backed up by a secure lineage, they were among the highest noble titles in Russia during Dashkova's time. However, no Russian prince or princess was ever in line for the throne nor related to the royal family. The titles "grand duke" and "grand duchess" were used to indicate heirs to the throne and their spouses.

Thematic Essays

Arduous and Delicate Task:

Princess Dashkova, the Academy of Sciences, and the Taming of Natural Philosophy

MICHAEL D. GORDIN

*P*rincess Ekaterina Romanovna Dashkova has been a woman more often recalled than remembered. Born in St. Petersburg in the uncertain years that followed in the wake of the tumultuous reign of Peter the Great, Dashkova proved in the second half of the century to be an exceptional personage in an age when Russia was truly a boundary to Europe: a bastion of the Enlightenment with the most extensive slavery system in the world; a center of opulence amid a massive impoverished peasantry; and the stage for many of the century's most impressive women in a land of pious Orthodoxy and conservatism. Dashkova's own role in this period was visible and well documented, but historians, chroniclers, and the popular imagination—in Russia, where she is relatively well known, as opposed to the West, where she is almost unheard of—focus on two aspects of her career: the extent of her involvement in the coup that brought Catherine II (the Great) to the throne of Russia in June 1762; and her position as the head of both the Imperial Academy of Sciences and Arts in St. Petersburg and the Imperial Academy of the Russian Language, posts she held for over a decade.[1] She was the first female public appointee in Russia, and the first woman to head any learned society in the world. If this were not enough, she also penned (in French) autobiographical memoirs that stand as one of the great Russian literary achievements in the genre, notwithstanding the language of their composition and the fact that Dashkova spends most of the memoirs "hiding" her innermost thoughts from the reader.[2]

Dashkova stands out to observers today as a grand exception in the eighteenth century, and to a certain extent she appeared that way to her contemporaries as well. An Irish friend and companion of her later years, Catherine Wilmot, wrote a letter from Moscow to her sister Alicia on December 8, 1805, that set much of the tone for writing about Dashkova since:

> I have since I came here often thought what a task it would be to attempt to draw the Character of the Princess Dashkaw! I for my part think it would be absolutely impossible. Such are her peculiarities & inextricable varietys that the result would only appear like a Wisp of Human Contradictions. 'Tis the stuff we are all made of to be sure, but nevertheless nothing is more foreign from the thing itself than the raw materials of which it is made! And woe betide individuality the moment one begins to generalize. You will always conceive her a piece of perfection when you take my experience of her, just as you would suppose Europe a Paradise if you never lived out of Italy & judged of the rest accordingly. But she has as many Climates to her mind, as many Splinters of insulation, as many Oceans of agitated uncertainty, as many Etnas of destructive fire and as many Wild Wastes of blighted Cultivation as exists in any quarter of the Globe! For my part I think she would be most in her element at the *Helm of the State*, or Generalissimo of the Army, or Farmer General of the Empire. In fact she was born for business on a large scale which is not irreconcilable with the Life of a Woman who at 18 headed a Revolution & who for 12 years afterwards govern'd an Academy of Arts & Sciences.[3]

Already the deck has been stacked against remembering Dashkova in her full context, for she is painted primarily as a *woman*. No one could sanely deny that she was indeed a woman, but there are limitations that come with such a narrow scope. In the nineteenth century, for example, Dashkova was primarily analyzed with an emphasis on her involvement in the Catherinian coup, because politics was a sphere in which women had virtually no presence, and therefore Dashkova was an anomaly to be treated as such.[4] For instance, Aleksandr I. Gertsen (Herzen), Russia's great liberal socialist thinker, chided the Princess's erstwhile companion and chronicler for not focusing on this fact enough: "All of this is true, but Miss Wilmot forgets that, above all, Dashkova was born a *woman* and remained a *woman* her entire life."[5] The key to presenting her as a political actor for Romanov thinkers was to focus on her essential femininity and deduce her political activities from there.

In the twentieth century, on the other hand, when women's participation in politics (although not, perhaps, in military coups) had become more common, the historiography took a more determined focus on her involvement in the academies. To Dashkova's contemporaries as well, her natural philosophical engagement sparked wonderment in contrast to her political engagement (the latter being not nearly as remarkable for taking place under the most powerful female ruler in modern European history). A 1785 article in the *Edinburgh Magazine* praised "[t]he great mental powers of this extraordinary lady, her thorough knowledge of various sciences, and the vast acquisition she has made of every species of useful information, from the most eminent philosophers and learned men of this age, during her travels through the most polished and civilized nations of Europe, attracted the attention of her Imperial Majesty the Empress of Russia." To the Scottish author, Dashkova demonstrated "how much better suited the fair sex is often, not only for the highest employments of governing vast and extensive nations, of which Herself [Catherine] is the most illustrious example; but of directing the arduous and delicate task of the various speculative sciences and nobler arts."[6] Given the tremendous scientific advances made by women in the twentieth century, Dashkova has served for many as a reminder that women have long contributed to science, or rather, one should say, to natural philosophy and natural history—the Enlightenment terms for what would later emerge as the physical sciences and the life sciences, respectively.[7]

But what is wrong with that, we might ask, given that it was indeed extraordinary for a woman to be so engaged with the sciences? Primarily, it prevents us from considering Dashkova's academic service in her own terms—she quickly overcame any scruples that being a woman marked her as unfit or special in this sphere—and it leads to an overweening stress on who Dashkova *was* biologically rather than what she *did* historically. This essay is an effort in the latter. Two salient features mark her administration of the Academy of Sciences (as distinguished from her work at the Academy of the Russian Language, which concerned the codification of the Russian language): first, she was highly competent, coaxing the Academy from the brink of ruin into one of the most flourishing learned societies of the late Enlightenment; second, she almost never engaged with the actual content of the natural philosophy and natural history produced at the Academy, preferring to administer from a position of intellectual distance (the few cases when she did actually debate academicians on their studies are made much of by historians, precisely because they were so exceptional).[8]

These two features are quite noteworthy for an academy of sciences in the late eighteenth century; they make Dashkova a useful figure with which to trace a major conceptual and institutional transformation. In general, academies were run, often poorly, by one of their members, who was more or less engaged in some kind of active research at the time. Most academy directors burst their budgets, disintegrated into vituperative internecine disputes, offended the courts that supported them, and so on. Dashkova's tenure at the Academy of Sciences was a break in all of that. She stood at the very dawn of the metamorphosis of "natural philosophy" into "science," and one of the major hallmarks of that shift was the transformation of "science" into an affair conducted by professionals; that is, people whose task was to explore the natural world. These emerging professionals controlled a form of specialized knowledge, policed themselves, and fulfilled a social function responsibly. Their administration could be devolved outside of their ranks, to a bookkeeper or a government functionary. That is to say, by exploring the state of the Academy of Sciences in St. Petersburg when Dashkova entered it, and then how she managed it during her lengthy tenure as director, one can observe irreversible changes wrought onto the fabric of how natural inquiry was conducted—a separation of the conduct of investigations (the knowledge) from the "trappings" of administration (the enabling powers that are not knowledge of the natural world). This is the crucial transition from the Enlightenment picture of knowledge to the workaday life of modern science, and Dashkova presided over it.

Unworthy: Dashkova Comes to the Academy

Dashkova was born on March 17, 1743, to Marfa I. and Count Roman I. Vorontsov, a scion of a highly distinguished noble St. Petersburg family. Her childhood, as she recalled it, did not presage that of a gregarious and active woman of high society, politics, and public administration: "From my infancy I wanted to be loved, I wanted to interest everyone that I loved, and when, at the age of thirteen, I grew to perceive that I was not producing this effect, I became an isolated being . . . My eyes, attacked mostly by illness, did not allow me to read, for which I was, I could say, passionate. A profound melancholy, of reflections on myself and above all that to which I belong, changed my lively, cheerful, even malicious character: I became serious, studious, I spoke little and only with knowledge of the matter."[9] She had books read to her, and was particularly drawn to major Enlightenment thinkers like Voltaire and Montesquieu, the former of which she would eventually meet. This childhood—Europeanized, educated, and caught up in the trappings of the intellectual Enlightenment—was privileged and sheltered, but was not atypical for a member of her class and sex. She was raised, as were most such girls of this period, to be married to an appropriate spouse, and this happened in February 1759, when she wed Prince Mikhail I. Dashkov, from another ancient Russian princely line hailing from the Monomakh clan of Smolensk. This union endowed her with her honorific "princess," which she bore proudly until her death.

Her more or less happy marriage to Dashkov had two significant consequences for her. Since she had been raised in the rarified circles of St. Petersburg, young Ekaterina had never learned to speak Russian properly, using French almost exclusively. The Dashkovs, however, were a Moscow family who (although still highly Europeanized in fashion and conduct) held tightly to the Russian language and could not converse in French. Dashkova learned Russian to communicate with her in-laws, and through the rest of her life, her French memoirs notwithstanding, held to the primacy of the Russian language with the zeal of the convert.[10] The second important consequence of the marriage was the particular way her husband, who was a vice-colonel in the leib-guards, introduced her into specific higher circles of Russian political life associated with the military.[11] It would be these associations that would put her in closer contact with the prospective tsarina, and enable her to assist in the coup that brought Catherine to power.

Dashkova's public life can be fairly cleanly divided into four major periods. The first was from 1759 to May 1763; that is, from the moment Dashkova met the Grand Duchess Ekaterina Alekseevna to the cooling of relations after the latter became Empress Catherine II. Dashkova's main activity during this time was assisting in Catherine's ascension to the throne. The second period, from the summer of 1763 until the end of 1782, saw Dashkova without much influence in the court at St. Petersburg; she devoted the time to her children's education and to traveling around Western Europe as a means to that end (there were two trips, 1770–71 and 1775–82, both studded with luminaries of the French and Scottish Enlightenments). The third period, from 1783 until the fall of 1794, marks her tenures as head of the two academies and occupies the central focus of the present essay. During the final period, from the fall of 1794 until her death in January 1810, Dashkova was informally and then formally removed from all her public posts and intervened in public life only episodically.[12] It is from this last period that we have her memoirs and the Wilmot sisters' recollections of her. Hence all of it is retrospective and is of limited assistance in gleaning what Dashkova thought or felt while in the midst of her activities. To examine her role at the Academy of Sciences as it happened, we have to turn to contemporary documents, many of which have recently been plumbed from the depths of the Russian archives.

Perhaps the most astonishing aspect of Dashkova's directorship of the Academy of Sciences was that she managed to be named to this post in the first place. As she recalled it in her memoirs, the whole affair started as a whim of the empress at a gala ball. Catherine drew her aside for a private audience and put forth her proposition. Dashkova wrote decades later:

> I was flabbergasted when Her Majesty told me that she wanted to offer me the place of the director of the Academy of Sciences.

My astonishment deprived me of speech, and the Empress had time
to tell me several very flattering things and that she believed would
encourage me. "No, Madame," I said, "I cannot accept a post above
my capacity. If Your Majesty is not making fun of me, I would tell her
that from attachment to her, amid many other reasons, I cannot rush
to render myself ridiculous and even culpable, of having made such a
choice!"[13]

Dashkova further recalled:

This conversation gave me a fever, and I believed that my entire
physiognomy was disorganized; for I perceived on the faces of some
of my companions (near to those I returned to sit by), the satisfaction
that they had, believing that it was a disagreeable scene for me that
had taken place. . . . I wrote immediately upon my return home a
letter which might have annoyed a different sovereign; because I
allowed myself to tell her that sometimes the private life of a monarch
narrowly escapes the pen of history, but never the bad or harmful
choices she makes; that God himself, in making me a woman, had
excused me from the post of the director of the Academy of Sciences;
that I knew myself for an ignoramus and that I had never angled for
being incorporated into a learned society, even in that of Arcadia,
where for a few ducats, in Rome, I could have achieved the honor of
joining it.[14]

Dashkova's letter, whose exact date is uncertain but was penned no later than January
24, 1783, confirms her retrospective recollections of an initial refusal that was grounded
in a sense of her own "unworthiness" as a woman, a fiercely gendered response: "But,
sovereign, my capacities are weaker than my zeal, and if your highness has finally
decided upon what I heard from you, then I beg you to edify and direct me, to indulge
me and not to offend me with the assumption as if I deserved this distinguished post,
which, in my opinion, does not belong with my sex."[15] Obviously, despite her early
protestations, she relented, largely because of a conversation she had with Catherine's
long-term favorite, Grigorii A. Potiomkin (Potemkin), who convinced her that
Catherine was serious and should not be disobeyed on this matter.

It seems that to all concerned (especially Dashkova), the appointment of a woman
to direct the Academy of Sciences, which had been founded by Peter the Great as part
of his reform project to modernize Muscovite Russia, was unthinkable.[16] This raises
the very legitimate question as to why Catherine would do something that was so
plainly beyond the pale of both custom and common sense. The surface reason was

that Dashkova was a woman, the two of them were friendly and respected each other intellectually, and Catherine wanted to demonstrate that she thought there was no reason why women could not serve in any public post in Russia. This proto-feminist interpretation is mostly superficial, however. The reasons to appoint Dashkova were not philosophical but political, both foreign and domestic. In terms of Catherine's "foreign policy" of Enlightenment, her willingness to appoint a woman director of the Academy was a bold statement of her position as a leading advocate of the Enlightenment, daring to call the bluff posed by the rhetoric of sexual equality of many contemporary thinkers. (The French Academy of Sciences, by contrast, would not admit women as members until after the Second World War.) The domestic policy reasons, however, were probably more salient to Catherine's reasoning. Dashkova was an intelligent and capable woman, skilled in political intrigue, and she was back in St. Petersburg; better to appoint her to the political backwater of the Academy, where she would be isolated from court machinations, and to hold her close but neutralized.[17]

Catherine knew that Dashkova would be quite busy at the Academy if she were to take her post at all seriously, as she anticipated she would. The Academy of Sciences, so optimistically staffed with European scholars and savants at the end of Peter's reign, had somewhat stagnated in its influence domestically, although the return in 1766 of Leonhard Euler, the most prominent mathematician of the century, from his hiatus at the Berlin Academy of Sciences ensured that its influence abroad was still strong. Technically, the president of the Academy from 1746 until 1798, encompassing within its copious span the entire reign of Catherine, was Kirill G. Razumovskii, but he paid absolutely no attention to the day-to-day affairs, and Catherine decided to create a post of director to manage the unraveling Olympus of the Russian Enlightenment. The first director she appointed (on October 25, 1773) was Vasilii G. Orlov, brother of Catherine's sometime lover and Dashkova's antagonist Grigorii G. Orlov, who masterminded the coup against Catherine's husband, Peter III (Dashkova's assistance in that event merely worsened their already-fraught relations). Orlov did not take to the labor of the Academy and retired on December 15, 1774.[18] In his absence the acting director was the writer Aleksei A. Rzhevskii, who had been former vice-director, but he was unable to serve during the crisis of the Emel'ian I. Pugachev serf uprising and was soon replaced by Sergei G. Domashnev, who was appointed on July 22, 1775. Domashnev gave his opening speech on December 11, and held the post until late 1782.[19]

Domashnev was an unmitigated, colossal catastrophe. He ruled with arbitrary, highly centralized, almost despotic personal power, asserting authority over almost every decision that had been customarily left to the academicians, including the approval of appointments.[20] On January 10, 1780, for example, he declared that academicians should publicly explain why so many of them had failed to conduct less research than previously proposed. The academicians felt insulted, claiming that they could not all be as productive as the preternaturally gifted Euler.[21] Active scholarly work almost completely collapsed. Domashnev prevented the publication of many works, almost never attended the regular meetings of the body, failed to fill (or even attempt to fill) many vacated posts, embezzled funds from the Academy's treasury, withheld incoming books from the library for his own personal collection, and allowed the Academy's *gymnasium* for educating young Russians to fall into disrepair.[22] Needless to say, Domashnev was not very popular and, in an unprecedented move, the academicians openly called for his removal. The conflict came to a head with Domashnev's dismissal of academician Semion K. Kotel'nikov, in April 1782, and his subsequent attempt to transfer the latter's cabinet of natural history to another academician, Peter Simon Pallas. The academicians claimed that "Such an authoritative act would overturn the entire system of academic obligations, and each academician in the future would begin to fear a similar unkindness, if his duty depended only on the director." Insisting that the dismissal violated the Academy's 1747 charter, they asked Domashnev to retract his decision. He ignored their protest—including their public statement of no confidence on August 22—and the affair ended with Domashnev being shown the door by Catherine.[23]

The Kotel'nikov affair gave the St. Petersburg court some pause. These academicians were traditionally expected to be pliant servitors of the government, answering technical queries and raising the state's prestige, not fomenting rebellion. The responsiveness by the government to the scholars' complaints has been taken by at least one historian as evidence that the government learned that it could not impose its will arbitrarily on the academicians.[24] Not quite. True, academic reform had been vigorously pursued as a topic of discussion during Domashnev's rampage through the system, even by the eminent mathematician Euler himself. Ever loyal to his administrative obligations on the Academy's governing council even while he was overhauling the foundations of physics and mathematics, Euler worried extensively over how Domashnev's arbitrariness was bankrupting the Academy. He proposed that the membership of the Academy be sharply curtailed, and that sales of newspapers, books, and calendars be expanded (the Academy controlled the printing industry in eighteenth-century Russia). He also thought the administration should be undertaken by someone of high status who could grease the wheels of power with the tsar (or tsarina) in case of difficulty. An academic commission of February 19, 1767, further expanded on these thoughts, angling for an expansion of academic privileges, such as an elimination of censorship on imported books and a raise in salary.[25] This commission's charter was not approved by the empress, however, and its failure replayed a familiar pattern of Catherine's reign; that of toying with liberalizing reforms and then flinching at the last moment.

But Catherine and her courtiers did learn something from the Domashnev years and especially from the circumstances of his removal. If the Academy were not properly governed, it could turn into a seed of instability that could disgrace the state and alienate the elite. In other words, the institution needed to be administered competently or the scholars would get uppity. Dashkova's appointment was not an answer to the call for reform; rather, she was a spoiler sent in to prevent the academicians' calls for deeper, more far-reaching reform to bolster their autonomy.

Episodes from the End of the Enlightenment: Dashkova's Tenure

Dashkova's career very much partook of the Enlightenment. She was deeply engaged with the ideas of her time, but paradoxically her legacy became one of the signal moments of the end of the Enlightenment in its Russian incarnation. By the "end of the Enlightenment" here, I mean those aspects of Dashkova's administration of the Academy of Sciences that contributed to the decline of a particular eighteenth-century vision of the pursuit of knowledge and the birth of professionalized science in the nineteenth century. These transformations were unplanned, and they stemmed from expressing rather straightforwardly some of the most basic features that characterized Enlightenment learned institutions. By institutionalizing them effectively in the late Catherinian context, Dashkova began to purge the pursuit of natural knowledge—including its research, publication, and distribution—in Russia of some of its more distinctive Enlightenment characteristics.

Episode 1. Euler and Inauguration: Enlightenment Knowledge and the Court
One of the most salient characteristics of natural philosophy and natural history practiced in scientific academies in the eighteenth century, as opposed to those activities in the universities or in private salons, was the close connection among those institutions, the knowledge they produced, and the court.[26] Dashkova's first instinct when she received her appointment was to perpetuate this close relationship; unintentionally, however, she ended up generating more distance, leading to the ostensible separation of science and politics that characterizes the modern period.

Dashkova knew that her appointment to the Academy would be controversial among the academicians—who were still energized by the drubbing they had given Domashnev—and she sought to extinguish any rancor by acting according to the refined etiquette protocols she knew best. As she narrated in her memoirs, her first action at the Academy before assuming her appointment was to approach Euler, the most distinguished scholarly (although not by rank) member of the Academy, and ask him to repay the kindness of her visit with his approval. She took the completely blind old man to her inaugural assembly in her coach: "Then I entered into the hall of sciences, I said to the professors and adjuncts that were assembled there that as testimony to the respect which I had to the sciences and enlightenment, however unfortunately ignorant I was myself, I was unable to find a more solemn way of proving this than by having myself introduced by Monsieur Euler."[27] (Within that very year, Euler was to pass away, and Dashkova respected the service his silent protection had rendered her by memorializing him in at least two special assemblies on September 11 and October 23.)[28] Her speech to the Academy on her inauguration, on January 30, 1783, was recorded in the Academy's minutes as follows:

> I dare to assure you, sirs, that the choice that her Imperial Highness made of my person, having laid upon me the presidency of this assembly, is for me an infinite honor, and I beg you to believe that these are by no means empty words, but a feeling by which I am deeply touched. I am ready to agree with the fact that I am inferior in enlightenment and capacities to my predecessors in this post, but I am not inferior to anyone of them in the directness of my own advantages, which will always instill in me an obligation to defer what is due to your talents, sirs. Far from ascribing to myself your merits, I regard it as my obligation to inform her Highness about the merits of each of you in particular with respect to the utility which the entire staff of the Academy brings to the good of her empire. This is the single advantage which I can promise you from my appointment; but since this will be exclusively in the care of your interests, then I hope that I, that my activity, based on this principle, will be able to evoke among you, sirs, a competition, in which each of you, working for the sake of your own fame, will not spare either energy or labor and that, in the end, thanks to our united efforts, the sciences will not from now own exist fruitlessly on our soil here; but, having settled here, they will leave deep roots and will blossom, finding themselves under the protection of a great monarchy which values learning.[29]

Through her instincts of flattery and ceremony, therefore, Dashkova situated herself as the conduit between Academy and court, and, more personally, between individual scholar and empress. But this inadvertently changed the function of the Academy of Sciences, subtly but decisively: now the Academy was understood as a place where learning was shielded from turmoil and friendly competition was provoked for the sake of learning. It was *not* a collection of hired factotums to amuse the empress. Dashkova had managed to use the tropes of courtly etiquette to move the Academy from an adornment of the court to a servitor of Russia.

Immediately upon assuming office, she instituted a series of reforms meant to regularize the Academy's position in the civil service, but also to render it distinct. On February 3, 1783, she proposed that all academic employees wear a newly commissioned uniform made of purple drape, with light yellow piping, and a cravat of light green, to be designed by Peter Pallas and approved by Dashkova herself.[30] This suggestion had the twin effect of making the academicians feel integrated into the state civil apparatus, as opposed to the court, but also to be distinct to the eye from other servitors. Science was, in other words, becoming a profession.

Episode 2. Financial Reform: The Bureaucratization of Knowledge

Notwithstanding repeated invocations of the otherwordly disinterestedness of science and natural philosophy, remarkably little knowledge of nature can be produced without substantial (or at least stable) finances. Dashkova's success in this realm was striking. When she arrived at the Academy of Sciences, it was in complete disarray. When she left her active administration of the institution in late 1794, the Academy had 100,000 rubles in the bank and 46,592 in cash; the bookstore and library were 390,000 rubles in the black, and the value of the assets at the printing press was 23,000 rubles—all of this after taking into account 100,000 rubles sunk into a new building. In sum, this was over half a million rubles of earned money.[31] The obvious way for an Enlightenment academic administrator to solve the financial arrears of his (or, in this case, her) institution was to ask the monarch for funds. Despite pledges of aid from Catherine, this was not what Dashkova chose to do. Instead, she reformed the Academy's internal finances with the zest of the most fastidious merchant. She ran the Academy like a business, and ran it well—another feature of her tenure that marked a shift away from Enlightenment practices and foreshadowed the rise of a more "bureaucratic" conception of administration, one that was more routine, impersonal, and efficacious.

This process of restructuring the Academy of Sciences so that it ran more like an efficient boarding house or press than like a cantankerous collection of savants and courtiers began as soon as the ceremonies of Dashkova's inauguration were concluded. She immediately had the papers of the administrative chancellery sent to her. "It was in reading these," she recalled, "that I was able to grasp, at least in part, the understanding of the task which I had to fulfill." The gravest problem was economic: the Academy was in debt to bookstores in Russia, Paris, and Holland. To compensate for the shortfall, instead of asking Catherine for succor, she raised the prices of Academy publications thirty percent, a shrewd business move that soon made substantial inroads in the Academy's debts.[32] She also reinvigorated the *gymnasium* from the brink of ruin. Although the Charter of 1747 demanded that there be fifty students enrolled, there were only twenty-seven in attendance at her inauguration, nine of whom could not attend classes due to complete ignorance of the sciences. Within three years, through a series of important pedagogical reforms, Dashkova managed to preside over eighty-nine competent students.[33]

In 1786, Dashkova wrote (in French) a lengthy report to Catherine detailing the transformations she had wrought on the Academy, all in the view of making it a more efficient bureaucratic institution. This document consists of a series of forty-five numbered problem areas, with an indented comment under each in smaller print explaining the reforms she had undertaken to remedy them. The very first point concerned finances: "1. Monetary affairs were extremely muddled as a consequence of the negligence with which so-called permanent sums were delimited from economic (special) sums; permanent [sums] included money allotted annually by the treasury for the maintenance of the Academy; special sums were comprised of money obtained from the sale of books and other economies. These sums turned out to be conflated with each other."[34] She alerted Catherine to the fact that she had begun reprinting the transactions (*Commentarii Academiae*) of the Academy (no. 8), which had been left unpublished since 1779—this lapse in publication had seriously cut into the Academy's revenues from their sale abroad—and adds as an aside (no. 9) that she had done something for scholarship as well:

> [Before] The professors, burdened with matters foreign to their sciences, did not have time to study their specialties, which hurt the successes of science.

> [Now] Each of them can study their science completely freely, not meeting any obstacles from my side; with all affairs they turn directly to me and receive a speedy resolution, not submitting to chancellery rigmarole, which frightens some of them.[35]

It almost seems that Dashkova had no interest as to what "their science" actually consisted in, as long as they could do it in the most efficient manner possible. (Of course, it was not all smooth sailing, as when she butted heads with the academicians in September 1788 by micromanaging their expenses for learned correspondence; she eventually retracted her interference.)[36]

Dashkova also actively tried to raise the prominence of the Academy. She knew that by appointing more honorary and foreign members, she could increase the prestige of the Academy at essentially no cost, and she appointed forty-seven honorary and foreign members (including Benjamin Franklin), a total amassed in barely over a decade that comprised a full twenty-five percent of all honorary academicians appointed during the entire eighteenth century.[37] She also instituted a number of moneymaking enterprises that would transform the finances of the Academy of Sciences, beginning with restructuring the book trade in Russia by organizing the Academy's press under more efficient principles.[38] The resulting funds were then channeled into material improvements in the laboratories and botanical gardens of the Academy, and then finally into the long-discussed construction of new quarters—a neoclassical façade that still houses Academy facilities today.[39] Interestingly, in the fall of 1783 the reason she gave Catherine for actually undertaking the building was that the Academy needed better space for selling books, for hosting (profit-making) public lectures, and then finally for housing the Academy's servants.[40] All in all, Dashkova's Academy ceased to be merely an appendage of the court designed mostly for show; it was a functioning business enterprise. It would be decades before other European academies took the same approach to their administration.

Episode 3. The Republic of Letters: Knowledge and Fatherland

In the nineteenth century, science would be marked by sporadic but fiery debates over the role of nationality in the production of knowledge. Science became an issue of national prestige: some nations were deemed (by their patriots) to be "more scientific" than others. In this area, too, Dashkova's tenure was a harbinger of a future era. A hallmark of her reforms was the promotion of nationality and language in the institution's affairs—a move towards the more nationalistic, nineteenth-century modus operandi.

By contrast, one of the most characteristic features of Enlightenment knowledge production was the so-called Republic of Letters, the continental and even intercontinental exchange of correspondence for collaborative resolution of heated disputes in philosophy, politics, and belles lettres. This exchange was not so much "international," since there were few well-defined nation-states for most of the eighteenth century, but rather "non-national"; and since almost everyone corresponded in French, it was functionally universal. One of the reasons Dashkova was appointed to direct the Academy was because she was already familiar with many of the major foreign denizens of the Republic of Letters.

During her travels in Europe, which were strongly motivated by her desire to inspect the Enlightenment bastion of Edinburgh University as a place for educating her son, Dashkova met with such luminaries of the epoch as Voltaire and Denis Diderot, meetings she dwelt on in her memoirs.[41] Although she and Catherine were both in some sense products of the Western European Enlightenment, that Enlightenment in fact came in many varieties. Catherine was a resolute Francophile, as can be seen in her correspondence with Voltaire, her copious quotations from Montesquieu in her *Instruction de Catherine II* (the *Nakaz*),[42] and her soliciting of Diderot to come to St. Petersburg. Dashkova, on the other hand, was partial to the Scottish and English Enlightenments, which tended to be gradualist and conservative in the sense championed by Edmund Burke: small changes grounded in tradition, all within the context of a protective monarchy. John Parkinson, a British visitor to St. Petersburg, recalled a conversation with Dashkova from November 1792: "We did not get away till near two after holding a long conversation with the Princess Dashkoff who was excessively civil. Though she blames us for some things, upon the whole she is a great admirer of the English Nation, envies us our constitution and regards a well-educated English Gentleman as the glory and perfection of his species."[43] Dashkova's attitude toward England provides a clue for how she acted at the Academy: she wished to run it like an English (or Scottish) university, not like a French debating society.

It may seem counterintuitive, but Dashkova's 1789 nomination of Benjamin Franklin as a foreign member of the Academy of Sciences, and his nomination of her as the first female member of the American Philosophical Society (APS), was in many ways tinged by this Anglophilia. (Dashkova would also propose APS member John Churchman as a foreign member in 1792.) The APS was modeled in part on the Royal Society of London, and Dashkova apparently perceived the American colonies as still very much of the English cultural sphere, and thus not tainted by French extremism (the American Revolution notwithstanding).[44] Dashkova herself was quite taken with Franklin when she met him during her second set of travels around Europe. At the time Franklin was serving diplomatically in France for the cause of American independence. On January 26, 1781, Franklin traveled from his home in Passy through Paris to the Hôtel de la Chine specifically to meet Dashkova, but she was not in and he had to return a week later. On Saturday, February 3, the two finally met.[45] Dashkova, in her memoirs, reveals Franklin only in a footnote:

> He had also such affection and esteem for me that he proposed me as
> a member of that respected and even celebrated Philosophical Society
> of Philadelphia; I was admitted there unanimously; I even received a
> diploma; after a certain period the Society did not miss a single occasion
> to send me the works which it had published. This packet contained a
> few [of them], as well as a letter of the secretary. That of Franklin flattered
> me more than that of the Duke [of Sudermanie], because I regarded him
> as a superior man and who combined with profound enlightenment
> a simplicity in his entire appearance and manners, and who with an
> unaffected modesty had a great deal of indulgence for others. I wrote to
> Franklin and to the secretary of the Philosophical Society, and thanked
> them sincerely for the works which they had sent me.[46]

Her reciprocal nomination of him was similarly good-natured. Franklin was an uncontroversial choice for the Academy, as he had been well known in Russia for years before his nomination for his seminal experiments and theories on electricity. His name was first mentioned in the Russian press in 1752, citing his work on atmospheric electricity, and he was clearly aware of the similar work of the Russian academician Franz Aepinus (and also, apparently, marginally aware of the work of the native Russian natural philosopher and academician, Mikhail V. Lomonosov).[47] Franklin was accepted unanimously on November 2, 1789.[48]

Dashkova's contacts with foreign natural philosophers and *philosophes* are strong markers of her active participation in the Republic of Letters. In her administration of the Academy of Sciences, however, one finds equal indications of a rising nationalism in her attitude toward scholarship that counteracted this earlier ecumenism. In her lengthy 1786 report to Catherine about her reforms at the Academy, she remarked:

> 41. Observations and discoveries conducted inside the country, were
> communicated abroad before their publication in Russia and, to the
> Academy's shame, were used by them before they were here.

I had the journal improved, so that the academicians do not have from now on to communicate such discoveries abroad until the Academy has gleaned from them fame for itself by means of print and until the government has used them.[49]

On 16 October 1786 she officially proclaimed that all Academy work would be published in Russia first, even though most of the academicians were still of Western or Central European origin and the language used in the *Commentarii Academiae* was still Latin. This stricture was quite a striking change in attitude toward the Academy, although it seemed to pass without a murmur from the academicians. From even before the age of Peter the Great, in early academies the scholars were not perceived as being part of and thus beholden to the national body; rather, their prestige abroad was intended to reflect on the glory of the monarch who was their patron. In that way of thinking, publishing abroad was essential, and whether or not one published at home was a matter of preference. Under Dashkova, the Academy began to be about serving the nation first in a notional, non-consultative way, and about becoming a crown jewel of the Romanov absolutist state as it evolved towards a nation-state. Dashkova's proposal for and administration of the Academy of the Russian Language—the new literary academy which was created to purify the Russian language, codify usage, and compile a dictionary—served the same function of early nation-building, although it cannot be discussed in detail here.[50]

Dashkova would often, in fact, claim that the sciences were native to Russia, and that the insistence on publishing in Russia first was merely a return to a time when Russia preserved knowledge for Europe during the imagined Dark Ages—a Slavic variant of how the Irish saved civilization. Dashkova wrote in her memoirs: "The sciences had been transported from Greece to Kiev long before they went to several European countries, who accord so freely to Russians the name of barbarians. The philosophy of Newton was taught in these schools when the Catholic priesthood did not permit that it be brought into France."[51] Dashkova thus worked to build an image of the Russian nation as a scientific state. She took the initiative in 1784–1787, for instance, to publish the works of Mikhail V. Lomonosov—the first Russian national to be appointed academician and often touted by contemporaries and historians since as the founder of science in Russia—a symbol of Russian ingenuity and parity with other nations then congealing in the West.[52]

Exits, Graceful and Not: Dashkova's Retirement and Legacy

Dashkova managed in just over a decade to set in motion a series of reforms that would not just tame natural philosophy and the practices of Enlightenment in Russia, but would resonate with measures taken across Europe to move towards pan-continental regimes of professionalized, bureaucratized, and nationalized science in the century to come. Dashkova's tenure at the Academy was so amicable that the radical nature of these transformations was not remarked upon explicitly at the time. It is therefore surprising to find that her manner of leaving was colored by revolution, rupture, and rancor, although it provides a historically satisfying closure to the transformative nature of her efforts at the Academy.

Most historians date the beginning of the end of Dashkova's directorship of the Academy of Sciences to the November 1793 publication by the Academy's presses of Iakov B. Kniazhin's play *Vadim Novgorodskii* [Vadim of Novgorod], a theatrical piece that made references to the public assembly (*veche*) of the medieval Russian city of Novgorod. According to many historians, Catherine—who was transformed by the French Revolution (and particularly by the execution of Louis XVI) from a polestar of Enlightenment to a *bête noir* of political reaction—took offense to what she interpreted as a play sympathetic to republicanism. And because Dashkova was responsible for its publication as head of the Academy, Catherine assumed she supported such views and punished her.[53] This account correctly reports the events in question but mishandles their interpretation. At no point could Catherine have rationally (or even irrationally) believed that Dashkova was actually a republican. Outside of the very real rise in the empress's paranoia about conspiracies at home and abroad against her reign, perhaps in favor of her son—who would, at her death in 1796, become Tsar Paul I—the political narrative of Dashkova's fall is more subtle than Kniazhin's eminently forgettable play.

The main point to remember about Dashkova's dismissal is that the academicians had nothing to do with it. They were happy with her and seemed genuinely sorry to see her leave. She had, after all, been running the Academy very competently for many years, and people seemed to have forgotten the days when the institution had been run like Domashnev's personal fiefdom and almost closed its doors. This selective amnesia played a role in Dashkova's exit, since the primary reason Catherine was upset with her former favorite for allowing a potentially subversive text to be published was not because she thought Dashkova was seditious, but that she had forgotten her responsibility as the Academy's censor. Dashkova had been so tremendously competent for so long, that this lapse, perhaps forgivable in less politically sensitive times, proved hard to ignore.[54] In addition, her favorite brother, Aleksandr R. Vorontsov, had recently been "retired" from government service after his protégé, Aleksandr N. Radishchev published the sharply critical *Puteshestvie iz Peterburga v Moskvu* [*Journey from St. Petersburg to Moscow*].[55] Dashkova suffered from the association (however loose) with both these objectionable events.

She was not subjected to the humiliation of a public dismissal, however. In August 1794, whether by hint from Catherine or her minions or, more probably, from an acute sense of which way the wind was blowing, Dashkova asked for a leave of absence from the Academy to tend to her health and domestic affairs. At the same time, she wrote the empress's state-secretary D. P. Troshchinskii that she wished to remain the fully active president of the Academy of the Russian Language, since she could perform those duties in the countryside away from St. Petersburg. On August 10, 1794, Troshchinskii informed her that she would be released for two years, and he accepted her suggestion of her relation Pavel P. Bakunin as a fill-in for her role at the Academy of Sciences. On August 24, 1796, she asked for an extension of her leave for another year (granted by Catherine on September 11, one of the empress's final acts). Dashkova thus remained the titular director of the Academy until after Catherine's death. On November 18, however, Catherine's son Paul fired her unceremoniously from both academies.[56] She retired to a state of relative poverty and a brief exile (whether voluntary or involuntary) to Korotovo, Novgorod Province, a place to which she never became accustomed and the resentment of which never left her even after her partial rehabilitation by Paul's son, and Catherine's grandson, Alexander I.

Dashkova lived through these final dark years with bitter recollections of the glory days at Catherine's side and then at the head of the Academy. Throughout all her travels of these last years, she continued to think of herself as a herald of the ideas and institutions that Catherine ostensibly stood for, and refrained from criticizing too sharply the monarch who treated her shabbily at the end. The academies she directed so well were remembered fondly, but it was clear that she never did realize quite how transformative her leadership there was. Even while penning her memoirs, which she saw as her central legacy, she did not perceive her tenure at the Academy as the dawn of a modern vision of science, nor did she see it as the end of the Enlightenment, whose creature she most definitely was.

Notes:

1. The two academies often cause confusion in terminology. The Imperial Academy of Sciences and Arts in St. Petersburg was established by Peter the Great in 1725 as a typical Western European learned society, complete with salaried academicians who were to pursue their own research and consult with the state on technical matters. The Imperial Academy of the Russian Language was founded under Catherine the Great at Dashkova's behest as a linguistic institution to standardize the Russian language. In the nineteenth century the Russian Academy was folded into the Academy of Sciences to become its second division on Slavic history and philology.

2. Princesse Dachkova [E. R. Dashkova], *Mon histoire: Mémoires d'une femme de lettres russe à l'époque des lumières*, eds. Alexander Woronzoff-Dashkoff, Catherine Le Gouis, and Catherine Woronzoff-Dashkoff (Paris, 1999); translated into English as E. R. Dashkova, *The Memoirs of Princess Dashkova*, tr. and ed. Kyril Fitzlyon (Durham, 1995). All subsequent references to the memoirs are from the French version in my own translation. On the convoluted publication history of the memoirs, see the editor's introduction, in E. R. Dashkova, *Zapiski. Pis'ma sester M. i K. Vil'mot iz Rossii*, ed. S. S. Dmitriev, comp. G. A. Veselaia (Moscow, 1987), p. 29; and the helpful E. L. Rudnitskaia, ed., *Spravochnyi tom k zapiskam E. R. Dashkovoi, Ekateriny II, I. V. Lopukhina* (Moscow, 1992). On memoirs as a genre of women's writing, see A. Woronzoff-Dashkoff, "Disguise and Gender in Princess Dashkova's *Memoirs*," *Canadian Slavonic Papers*, vol. 33 (1991), pp. 62–74; Catriona Kelly, "Sappho, Corinna, and Niobe: Genres and Personae in Russian Women's Writing, 1760–1820," in *A History of Women's Writing in Russia*, eds. Adele Marie Barker and Jehanne M. Gheith (Cambridge, 2002), pp. 37–61; Judith Vowles, "The 'Feminization' of Russian Literature: Women, Language, and Literature in Eighteenth-Century Russia," in *Women Writers in Russian Literature*, eds. Toby W. Clyman and Diana Greene (Westport, Conn., 1994), pp. 35–60, 43; Kelly Herold, "Russian Autobiographical Literature in French: Recovering a Memoiristic Tradition

(1770–1830)" (Ph.D. diss., University of California, Los Angeles, 1998), pp. 119–22, 208; and the seminal study by Barbara Heldt, *Terrible Perfection: Women and Russian Literature* (Bloomington, 1987), p. 71. On disputes about the authenticity of Dashkova's memoirs, see the analysis in M. M. Safonov, "Ekaterina malaia i ee 'Zapiski,'" in *Ekaterina Romanovna Dashkova: issledovaniia i materialy*, eds. A. I. Vorontsov-Dashkov et al. (St. Petersburg, 1996), pp. 13–22. For supplemental material not included in the initial publication, see V. G. Esipov, "K biografii kniagini E. R. Dashkovoi," *Istoricheskii vestnik*, vol. 9 (1882), pp. 668–75.

3. Martha and Catherine Wilmot, *The Russian Journals of Martha and Catherine Wilmot*, 1803–1808, eds. Marchioness of Londonderry and H. M. Hyde (London, 1934), p. 211.

4. Much of the recent scholarship on women and politics in the Russian Enlightenment points to the historical specificity of the debates about female rule, drawing from the path-breaking study by Brenda Meehan-Waters, "Catherine the Great and the Problem of Female Rule," *Russian Review*, vol. 34 (1975), pp. 293–307.

5. A. I. Gertsen [Herzen], "Kniaginia Ekaterina Romanovna Dashkova," in A. I. Gertsen, *Sobranie sochinenii*, vol. 12 (Moscow, 1957), pp. 361–422, 361.

6. "The Russian Academy," *The Edinburgh Magazine* (April 1785), pp. 304–7, 304.

7. D. Mordovtsev, *Russkiia zhenshchiny novago vremeni: Biograficheskie ocherki iz russkoi istorii*, vol. 2 (St. Petersburg, 1874), pp. 119–59; O. I. Eliseeva, "Ekaterina II i E. R. Dashkova: Fenomen zhenskoi druzhby v epokhu prosveshcheniia," in *E. R. Dashkova i A. S. Pushkin v istorii Rossii*, ed. L. V. Tychinina (Moscow, 2000), pp. 19–33; V. V. Ogarkov, *E. R. Dashkova: Eia zhizn' i obshchestvennaia deiatel'nost'* (St. Petersburg, 1893), p. 27; and Carol S. Nash, "Educating New Mothers: Women and the Enlightenment in Russia," *History of Education Quarterly*, vol. 21 (1981), pp. 301–16.

8. For example, in April 1786 she intervened with Academy corresponding member Johann-Michael Renovanz to finish organizing the cabinet of foreign minerals, insisting that he use the Linnaean system of mineralogical classification instead of Bergman's, since she thought the former was superior. L. V. Tychinina, *Velikaia rossiianka: zhizn' i deiatel'nost' kniagini Ekateriny Romanovny Dashkovoi* (Moscow, 2002), p. 143. The scholarship on Dashkova at the Academy is extensive but not always terribly illuminating. For some of the more prominent studies, see L. Ia. Lozinskaia, *Vo glave dvukh akademii*, 2d. ed. (Moscow, 1983); B. I. Krasnobaev, "Glava dvukh akademii," *Voprosy istorii*, no. 12 (1971), pp. 84–98; G. I. Smagina, "Akademiia nauk i narodnoe prosveshchenie v Rossii vo vtoroi polovine XVIII veka," *Voprosy istorii estestvoznaniia i tekhniki*, no. 1 (1991), pp. 39–46; idem, "E. R. Dashkova—direktor Peterburgskoi akademii nauk," in *E. R. Dashkova i ee vremia: issledovaniia i materialy*, eds. L. V. Tychinina et al. (Moscow, 1999), pp. 35–43, 36; G. A. Tishkin, "A Female Educationalist in the Age of Enlightenment: Princess Dashkova and the University of St. Petersburg," *History of Universities*, vol. 13 (1984), pp. 137–52; idem, "'Ee svetlost' madam direktor' (E. R. Dashkova i Peterburgskii universitet v 1783–1796 gg.)," in *Ekaterina Romanovna Dashkova*, pp. 80–93; M. P. Tolstoi, "E. R. Dashkova—organizator rossiiskoi nauki," *Vestnik Rossiiskoi akademii nauk*, vol. 63, no. 3 (1993), pp. 245–48; E. P. Ozhigova, "E. R. Dashkova—Direktor Peterburgskoi akademii nauk," in *Ekaterina Romanovna Dashkova*, pp. 94–102; Günther Schlegelberger, *Die Fürstin Daschkowa: Eine biographische Studie zur Geschichte Katharinas II.* (Berlin, 1935). H. Montgomery Hyde, *The Empress Catherine and Princess Dashkov* (London, 1935) is unreliable with respect to the facts of her involvement in the Academy.

9. Dashkova, *Mon histoire*, p. 15. Dashkova had a passion for books her entire life. The catalogue of her Moscow library ran fifty pages and contained almost 4,500 books on a huge variety of topics. See Tychinina, *Velikaia rossiianka*, p. 54; and A. Woronzoff-Dashkoff's essay in this volume.

10. M. I. Sukhomlinov, "Kniaginia E. R. Dashkova," in M. I. Sukhomlinov, *Istoriia Rossiiskoi akademii*, vol. 1 (St. Petersburg, 1874), pp. 20–57, 21.

11. Tychinina, *Velikaia rossiianka*, p. 31.

12. Tychinina, *Velikaia rossiianka*, pp. 91–92.

13. Dashkova, *Mon histoire*, p. 156.

14. Dashkova, *Mon histoire*, pp. 157–58.

15. Cited in E. R. Dashkova, *O smysle slova "vospitanie": sochineniia, pis'ma, dokumenty*, ed. G. I. Smagina (St. Petersburg, 2001), p. 272.

16. On the foundation of the Academy of Sciences by Peter the Great, see Michael D. Gordin, "The Importation of Being Earnest: The Early St. Petersburg Academy of Sciences," *Isis*, vol. 91 (2000), pp. 1–31.

17. Tychinina, *Velikaia rossiianka*, p. 130.

18. N. I. Nevskaia, ed., *Letopis' Rossiiskoi Akademii Nauk*, vol. 1 (St. Petersburg, 2000), pp. 607, 622.

19. Nevskaia, *Letopis' Rossiiskoi Akademii Nauk*, pp. 627, 629.

20. K. V. Ostrovitianov, ed., *Istoriia Akademii nauk SSSR*, vol. 1, 1724–1803 (Moscow, 1958), p. 320.

21. Nevskaia, *Letopis' Rossiiskoi Akademii Nauk*, p. 665.

22. Truth be told, the academicians had not been terribly well behaved under Orlov or Domashnev. On January 20, 1774, they had to be publicly reminded that they had to attend meetings regularly and publish, indicating that not all of them were as assiduous as they later claimed. See Nevskaia, *Letopis' Rossiiskoi Akademii Nauk*, p. 610.

23. Nevskaia, *Letopis' Rossiiskoi Akademii Nauk*, p. 688 (quotation), p. 693 (dismissal).

24. Ostrovitianov, *Istoriia Akademii nauk SSSR*, p. 321.

25. Ostrovitianov, *Istoriia Akademii nauk SSSR*, pp. 317–18.

26. On this, see Gordin, "The Importation of Being Earnest."

27. Dashkova, *Mon histoire*, p. 161.

28. Nevskaia, *Letopis' Rossiiskoi Akademii Nauk*, pp. 704–5.

29. Cited in Dashkova, *O smysle slova "vospitanie,"* p. 274. Dashkova came through on this promise. In 1784 the academicians suggested that any proposal by an adjunct had to be approved by a full academician, but Dashkova vetoed this because it diminished the fame that could be obtained by the adjuncts: "Fame is often the only reward for the labors of science: why encroach upon its attainment? Let everyone know about the discovery and judge it fairly; if there is some kind of scientific discovery communicated in the memoir of an adjunct, which has gone through the hands of a professor, then it is very possible to suggest external participation, which lowers the merit of the author, in particular, if it is a young scholar who has managed to incur envious ill-wishers for himself." Quoted in Tychinina, *Velikaia rossiianka*, p. 147.

30. Nevskaia, *Letopis' Rossiiskoi Akademii Nauk*, p. 699.

31. E. P. Ozhigova, "E. R. Dashkova—Direktor Peterburgskoi akademii nauk," in *Ekaterina Romanovna Dashkova*, pp. 94–102, 101–2. See also the account of the Academy's finances from Dashkova to Catherine, June 1, 1783; cited in Dashkova, *O smysle slova "vospitanie,"* pp. 282–87.

32. Dashkova, *Mon histoire*, p. 163.

33. G. A. Tishkin, "E. R. Dashkova i uchebnaia deiatel'nost' v Peterburgskoi akademii nauk," *Ocherki po istorii Leningradskogo universiteta*, vol. 6 (1989), pp. 190–207, 196.

34. Cited in Dashkova, *O smysle slova "vospitanie,"* p. 309. For the original French, see *Arkhiv Kniazia Vorontsova* (Moscow, 1881), vol. 21, pp. 389–402.

35. Cited in Dashkova, *O smysle slova "vospitanie,"* p. 311.

36. Nevskaia, *Letopis' Rossiiskoi Akademii Nauk*, p. 755.

37. G. I. Smagina, "E. R. Dashkova—direktor Peterburgskoi akademii nauk," in *E. R. Dashkova i ee vremia*, pp. 35–43, 39.

38. A. A. Zaitseva, "E. R. Dashkova i knizhnaia torgovlia Akademii nauk," in *Ekaterina Romanovna Dashkova*, pp. 110–27; and G. I. Smagina, "E. R. Dashkova i prosvetitel'skaia deiatel'nost' Akademii nauk," in *Ekaterina Romanovna Dashkova*, pp. 103–9. On publishing in eighteenth-century Russia in general, see Gary Marker, *Publishing, Printing, and the Origins of Intellectual Life in Russia, 1700–1800* (Princeton, 1985).

39. Tychinina, *Velikaia rossiianka*, p. 155.

40. Dashkova, *O smysle slova "vospitanie,"* p. 292.

41. Dashkova, *Mon histoire*, pp. 91–93 (Diderot), p. 99 (Voltaire). On her interactions with French thinkers, and how they perceived her, see A. Niv'er, "E. R. Dashkova i frantsuzskie filosofy prosveshcheniia Vol'ter i Didro," ed. M. I. Mikeshin, in *Ekaterina Romanovna Dashkova*, pp. 41–54; and V. A. Somov, "'Prezident trekh akademii': E. R. Dashkova vo frantsouskoi [sic] 'Rossike' kontsa XVIII veka," in *E. R. Dashkova i A. S. Pushkin v istorii Rossii*, pp. 39–53, 40. Dashkova's reform of the Academy's *gymnasium* was heavily based

on the model of Edinburgh University, as described in G. A. Tishkin, "'Ee svetlost' madam direktor' (E. R. Dashkova i Peterburgskii universitet v 1783–1796 gg.)," in *Ekaterina Romanovna Dashkova*, pp. 80–93, 84.

42. Catherine II's *Instruction* (*Nakaz* in Russian) was written for the Legislative Commission she assembled in 1767 to develop a new legal code for Russia.

43. John Parkinson, *A Tour of Russia, Siberia and the Crimea, 1792–1794*, ed. William Collier (London, 1971), p. 38. Dashkova was fluent in English, and her brothers Aleksandr and Semion actually served in the Russian diplomatic staff in England during Catherine's reign. In her visits to England, she was strongly taken by both Bath and its classic architecture and the climate of Oxford University, which she spoke of in terms that contrast sharply with her suspicion of the salons of Paris. On her Anglophilia and the way the English perceived her, see E. P. Zykova, "Anglofil'stvo E. R. Dashkovoi v kontekste russkoi kul'tury XVIII veka," in *E. R. Dashkova i rossiiskoe obshchestvo XVIII stoletiia*, eds. N. P. Karpichenko et al. (Moscow, 2001), pp. 89–106; and Anthony Cross, "Contemporary British Responses (1762–1810) to the Personality and Career of Princess Ekaterina Romanovna Dashkova," *Oxford Slavonic Papers*, n.s., vol. 27 (1994), pp. 41–61.

44. On the Royal Society as model for the American Philosophical Society (APS), see Edward C. Carter II, *"One Grand Pursuit": A Brief History of the American Philosophical Society's First 250 Years, 1743–1993* (Philadelphia, 1993), p. 14. On the early struggles of the APS, see Brooke Hindle, *The Pursuit of Science in Revolutionary America, 1735–1789* (Chapel Hill, 1956), chapter 7.

45. A. Woronzoff-Dashkoff, "Princess E. R. Dashkova: First Woman Member of the American Philosophical Society," *Proceedings of the American Philosophical Society*, vol. 140, no. 3 (1996), pp. 406–17.

46. Dashkova, *Mon histoire*, n. 179. See also N. N. Bolkhovitinov, *Rossiia otkryvaet Ameriku, 1732–1799* (Moscow, 1991), pp. 148–51.

47. Eufrosina Dvoichenko-Markoff, "Benjamin Franklin, the American Philosophical Society, and the Russian Academy of Science," *Proceedings of the American Philosophical Society*, vol. 91, no. 3 (1947), pp. 250–57. On the extent of early contacts between the APS and Russians, see idem, "The American Philosophical Society and Early Russian-American Relations," *Proceedings of the American Philosophical Society*, vol. 94, no. 6 (1950), pp. 549–610.

48. Nevskaia, *Letopis' Rossiiskoi Akademii Nauk*, p. 771.

49. Cited in Dashkova, *O smysle slova "vospitanie,"* p. 319.

50. On the Russian Academy and especially on its dictionary, see Sukhomlinov, "Kniaginia E. R. Dashkova" in *Istoriia Rossiiskoi: akademii;* M. Sh. Fainshtein, "E. R. Dashkova i 'Slovar' akademii rossiiskoi' (1783–1794)," in *E. R. Dashkova: Lichnost' i epokha*, eds. L.V. Tychinina and A.V. Semenova (Moscow, 2003), pp. 58–71; Ogarkov, *E. R. Dashkova*, p. 59; E. P. Chelyshev, "Ot Ekateriny Romanovny i Konstantina Romanova do nashikh dnei," *Vestnik Rossiiskoi akademii nauk*, vol. 63, no. 6 (1993), pp. 536–45; Tychinina, *Velikaia rossianka*, p. 145; and Hans Rogger, *National Consciousness in Eighteenth-Century Russia* (Cambridge, Mass., 1960), p. 118.

51. Dashkova, *Mon histoire*, p. 84. Parkinson in December 1792 indicated that this was a longstanding theme of Dashkova's: "The Princess Dashkoff says that during the extinction of the Arts in every part of the world for some time, when nobody could say what was become of them, at that time they took their abode in Russia" (Parkinson, *A Tour of Russia*, p. 50).

52. Ogarkov, *E. R. Dashkova*, p. 57.

53. This is the account offered, for example, in Ch. de Larivière, *Catherine II et la révolution française* (Paris, 1895).

54. Tychinina, *Velikaia rossiianka*, pp. 123–24.

55. Woronzoff-Dashkoff, "Princess E. R. Dashkova," p. 412.

56. Tychinina, *Velikaia rossiianka*, pp. 133–34.

Liberty Postponed:
Princess Dashkova and the Defense of Serfdom

MICHELLE LAMARCHE MARRESE

*F*or all the diversity that distinguished imperial Russia from the American colonies in the eighteenth century, both societies shared one crucial characteristic that set them apart from Western Europe: the presence of human bondage as a mainstay of economic power and social privilege. Since 1721, when Montesquieu launched his attack on the slave trade in *The Persian Letters*, Enlightenment thinkers had roundly criticized slavery on moral and utilitarian grounds. Slavery, they argued, was not only contrary to natural law, but hampered economic prosperity. The *philosophes* in France and Britain may have benefited indirectly from slave industries in the colonies, but—unlike their Russian and American counterparts— their economic well-being did not hinge on the labor of unfree servitors in their midst.[1]

By contrast, proponents of Enlightenment on the periphery of Europe could not escape the uneasy juxtaposition of ideals and experience in their everyday lives. To be sure, advocates of abolition could be found among America's founders. Other colonists, although aware of the inconsistency between principles and practice, maintained that unfree labor was a temporary evil that would pass away with the progress of human society. This was particularly true of plantation owners in the southern colonies. Most notoriously, Thomas Jefferson condemned slavery as an "abominable crime," yet remained a slave-owner throughout his life and even supported the expansion of slavery into the southwestern territories.[2] For the eighteenth-century Russian nobility and its rulers, however, proposals for serf emancipation comprised little more than an intellectual exercise, a subject for essay competitions and debate among officials, rather than a prelude to reform. Only the most enlightened members of the Russian nobility were prepared even to acknowledge the abuses inherent in serfdom and to allow for state intervention in the relationship between proprietors and their human property.[3]

Among the most eloquent supporters of serfdom in eighteenth-century Russia was Princess Ekaterina Romanovna Dashkova. In addition to her more celebrated role as director of the Imperial Academy of Sciences and Arts in St. Petersburg, throughout her life Princess Dashkova acted as the indefatigable manager of several landed estates. For Dashkova, as for other members of the ruling class, human bondage was central to the noble way of life: the princess relied overwhelmingly on serf labor for her income, for the care of her estates and children, and even as a source of entertainment and amusement.[4] Far from passing over this dimension of her life in silence, Dashkova's relation to her serfs was a recurring theme in her memoirs and her correspondence. Indeed, Dashkova's conviction that she epitomized the patriarchal ideal of the "good proprietor" (*dobryi barin*), devoted to the well-being of her dependants, was a significant element in her self-representation and long predated the writing of her memoirs.

The contrast between Dashkova as a leading purveyor of Enlightenment philosophy in Russia and her status as proprietress of thousands of human "souls" struck many of the princess's contemporaries and, subsequently, her chroniclers as incongruous. Both before and after the 1917 revolution, Russian historians vehemently criticized Dashkova for her defense of serfdom, labeling her one of those "false liberals who ardently demand freedom—for themselves."[5] In their denunciations of Dashkova, however, critics declined to consider the context of her apology for unfree labor, not to mention the similarities between Dashkova's views of the "lower classes" and the pronouncements of more renowned Enlightenment intellectuals. Indisputably, Dashkova's compatriots had their own interests at heart when they rejected the notion of serf emancipation as unviable in the economic and social conditions of eighteenth-century Russia. Yet Dashkova's espousal of serfdom, like that of her noble contemporaries, derived equally from her awareness of the state of corporate and individual rights in her native land.

Unlike its counterparts in Western Europe, the Russian nobility did not originate as a hereditary body but as the creation of the sovereign, who bestowed land and human labor upon his subjects in return for their service to the state. Indeed, before the reforms of Peter the Great early in the eighteenth century, the nobility did not exist as a distinct corporation. In light of the dependence of the elite on the monarch for its social and economic privileges, the status of the Russian nobility as a true "ruling class" remains the subject of debate. The eighteenth century, however, witnessed a protracted struggle on the part of Russian nobles to clarify their property rights in relation to other family members and to the state, and to bolster their corporate privileges vis-à-vis competing social groups. The nobility's campaign for corporate rights and greater liberty in relation to the sovereign was nonetheless founded on a significant paradox: the demand for the exclusive privilege to own and administer human property. In the words of one scholar, "The basic weakness of the [nobility's] position was that it sought to found its own liberty on the bondage of others."[6]

In their petitions to Catherine II on the eve of the 1767 Legislative Commission, the nobility unanimously argued that no other social group should enjoy the right to serf labor. Their efforts gradually met with success. Catherine's Charter to the Nobility in 1785 confirmed many of the privileges the nobility had garnered in previous decades: protection from arbitrary confiscation of property; freedom from corporal punishment; and, not the least, the exclusive right to serf ownership. At the same time, the Charter stopped short of guaranteeing the inviolability of noble status or property and defined the crimes meriting loss of rank and confiscation of property as broadly as possible. Thus, long after the promulgation of the Charter to the Nobility, nobles who indulged in deeds "irreconcilable with . . . noble dignity" risked forfeiting their status and their estates, and faced the prospect of corporal punishment and exile.[7] The enduring fragility of the nobility's security, in regard to both person and property, provides the necessary backdrop for Dashkova's defense of serfdom, which appears, at first glance, greatly at odds with her promotion of the Enlightenment in Russia. Despite her genuine interest in ameliorating the lot of the Russian peasantry, Dashkova's primary goal was to shore up the security and privileges of the ruling class.

Noblewomen and Estate Management in Imperial Russia

While Dashkova was unique, in both a European and a Russian context, as a woman holding public office, in her role as administrator of estates and serfs she was far from unusual among her female counterparts. The third daughter of Count Roman Vorontsov, Dashkova was born in 1743 into one of the most prominent families in eighteenth-century Russia. Her wealth, however, derived primarily from gifts from the sovereign and her savvy management of her husband's assets after his death, rather than from the Vorontsov family fortune. Upon her marriage to Prince Mikhail Dashkov in 1759, Dashkova received a dowry in moveable goods worth 12,917 rubles, as well as

10,000 rubles for the purchase of villages: far less than the one-fourteenth of her father's vast estate that Dashkova might have anticipated according to Russian inheritance law in regard to daughters.[8] Although the princess acquired a small plot of land and a house during her short marriage, her dowry was eventually expended on paying her husband's debts and the education of her children, rather than the purchase of villages.[9] It was only in the aftermath of her husband's premature death in 1764 that Dashkova embarked upon her career as estate manager.

The demise of Prince Dashkov marked the beginning of Dashkova's role as the proprietress of several thousand serfs. It also provided the catalyst for a prominent theme in Dashkova's personal mythology: that of her straitened circumstances in the years following her husband's death and her reluctance to accept the largess of Catherine II in return for her support during the 1762 palace coup that brought the latter to power. In her memoirs, Dashkova described the poverty she suffered as a widow in vivid detail and declared that she alone, in order to provide a sound inheritance for her children, built her considerable fortune through frugality and sacrifice. Not only in regard to her education, but also as a woman of considerable means, Dashkova never deviated from her self-portrait as a "self-made woman." Even during Dashkov's life, the princess announced her intention "not to incur any expense which might inconvenience my husband . . . I imposed the strictest economy on myself, wore nothing but the oldest clothes and remains of my trousseau, and restricted my *penchant* for books, etc., as much as possible."[10]

The discrepancy between Dashkova's lamentations about her financial affairs and evidence of the fortune she eventually amassed left the princess vulnerable to accusations of avarice and parsimony, first among her contemporaries and later among historians. Dashkova herself was well aware of these allegations and hotly defended herself against charges of greed in a letter to her friend Catherine Hamilton written toward the end of her life. For the sake of her children, Dashkova proclaimed, she did everything in her power not to touch their inheritance while settling her husband's debts, "leading a style of life and dressing beneath my station"—a motif she elaborated on frequent occasions in her memoirs.[11] Recounting the years she spent in Scotland, overseeing her son's education, Dashkova declared that, in light of this goal, "my own poverty affected me not at all."[12] Indeed, the princess devoted her first five years of widowhood to ordering her husband's financial affairs, handing over to his creditors much of their personal property, as well as "the few jewels" she possessed.[13] While maintaining she had been ignorant of the family finances during her marriage, Dashkova quickly demonstrated her business acumen and ability for estate management. She would later exhibit these same abilities to great effect in her role as director of the Academy of Sciences.

If Dashkova was unusual among her female peers as the recipient of financial generosity on the part of the sovereign,[14] in her ownership and active administration of her estates she acted well in keeping with Russian tradition. Property law in imperial

Russia offered unusually generous property rights to noblewomen. From 1753, not only widows, but also married women, enjoyed the right to alienate and administer their estates without their husbands' permission. Noblewomen's control of property was, moreover, not merely a legal formality. The elevation of Russian women's legal rights translated into a dramatic rise in female property ownership, culminating in women's control of as much as one-third of landed estates in private hands by the end of the eighteenth century—a state of affairs that persisted until the abolition of serfdom in 1861. Among these *pomeshchitsy*, or female proprietors, well over half of those who took part in property transactions from the mid-eighteenth century were married; approximately thirty percent, like Dashkova, were widows.[15]

As Dashkova, like her female compatriots, took on the responsibilities of estate administrator, she also assumed control of human labor, with all its attendant burdens and rewards. West European observers frequently singled out serfdom as the salient feature that separated "backward" Eastern Europe from the "enlightened" West.[16] In fact, the enserfment of the peasantry was a relatively late development in Russia, emerging in the sixteenth and seventeenth centuries in response to territorial expansion coupled with a severe shortage of agricultural labor. Moreover, by no means all of Russia's peasants were serfs: in the late eighteenth century, just over half of the peasant population was the property of individual proprietors. The remainder consisted of state peasants, who were administered by the government, and a smaller number of peasants who lived on estates belonging to the sovereign. While the latter were also subject to heavy financial burdens and military service, as well as constraints on their mobility, the lot of state peasants was considerably less onerous than that of serfs, who suffered closer supervision, not to mention abuse, at the hands of capricious proprietors or their bailiffs.[17] Historians continue to debate the fine distinctions between serfdom and slavery; overwhelmingly, however, they agree that by the mid-eighteenth century the powers of Russian noble proprietors over their human chattel were virtually unlimited. Certainly this was the view of European visitors to Russia, who commented that "the peasants in Russia . . . are in a state of abject slavery; and are reckoned the property of the nobles to whom they belong, as much as their dogs or horses."[18]

Although the right of serf ownership was confined to the nobility by the reign of Catherine II, the benefits of serf labor were by no means distributed equally among the elite. More than fifty percent of Russian serf owners in the eighteenth century possessed fewer than twenty-one peasants—a number that fell far short of the hundred serfs considered the minimum necessary for a noble family to eke out a precarious living on an estate. Another seventeen percent of the nobility made it over this poverty line, while a scant one percent qualified as grand seigneurs, or those who controlled more than one thousand souls. Thus, as the proprietress of well over five thousand serfs at the time of her death,[19] Dashkova was also exceptional among her peers, both male and female, in the sheer numbers of human property she controlled.

In assuming active control of her estates, Dashkova in no way transgressed accepted gender conventions for Russian women, either married or widowed. Although gender segregation was a prominent feature of family life in Russia as in the West, running the estate was a joint venture that demanded the attention of both husband and wife, or, in the case of widows, cooperation between mothers and sons. A traditional division of labor was undoubtedly the rule in many noble households. Yet the most striking feature of memoirs of estate life is that they describe noblewomen making investment decisions and supervising large-scale management as often as they portray women confined to tasks in the home. Lengthy absences of noblemen on state service, not to mention the control of estates by women such as Dashkova, ensured that Russian noblewomen, at some point in their lives, had to concern themselves as much with boundary disputes, recalcitrant serfs, and the price of wheat as they did with the clothing and education of their children. In short, the boundaries of permissible conduct for elite women, regardless of marital status, proved highly flexible in the realm of estate administration. Contemporaries praised noble proprietors of both sexes for acting as custodians of the family fortune, for saving the ancestral estate, and for passing on a sound inheritance to their children. Conversely, women who neglected their estates or managed them poorly were not forgiven on the grounds of their sex.

Princess Dashkova as Serf Owner

Throughout the memoirs, Dashkova's account of her life moves between her discussion of "public affairs" [20]— her role in the 1762 coup d'état that elevated Catherine II to the throne and her work as director of the Academy of Sciences—and her attention to her children and the prosperity of her peasants. To be sure, Dashkova was not alone in her obsessive scrutiny of every detail of estate affairs; noble family papers offer abundant evidence of female proprietors who rivaled Dashkova in their unflagging correspondence with bailiffs, their detailed accounts of estate production, and their efforts to improve the living conditions of their peasants. [21] Nonetheless, Dashkova's letters and memoirs offer unusually eloquent testimony on skilled estate management as a source of pleasure and identity for Russian noblewomen. Dashkova's contemporaries were unanimous in their acknowledgement of the princess as an exceptional manager. For some, such as the Wilmot sisters, Dashkova's labors on her estate inspired admiration. Catherine Wilmot exclaimed more than once at the range of the princess's responsibilities: "[S]he helps . . . the masons to build walls, she assists with her own hands in the making of roads, she feeds the cows . . . [S]he is a Doctor, an Apothecary, a Surgeon, a Farrier, a Carpenter, a Magistrate, a Lawyer . . . and yet appears as if she had her time a burthen on her hands." [22] For all their approbation of Dashkova's treatment of her peasants, however, the sisters were united in their conviction that serfdom was a corrupting influence on those who possessed human chattel and lay at the heart of the servility that permeated social relations at all levels of Russian society. In a moment of particular indignation, Catherine wrote to a friend, "I will take you down stairs into the Hall where dozens of Slaves are waiting with their offerings of Bread and Salt to greet the Princess! When she appears they fall down before her & kiss the ground with that senseless obeisance that stupefaction feels at the approach of superior Power! Her Lenity makes their Lot better perhaps than that of others, but that's saying very little for the System."[23]

Even the most judicious treatments of Dashkova's life betray similar unease, if not outright opprobrium, in their depiction of Dashkova's relationship with her serfs and her defense of serfdom. Several nineteenth-century historians took Dashkova to task for her attempts to make her estates yield as high a profit as possible. In a particularly critical account, V. V. Ogarkov belittled the princess for hoarding money in a manner befitting a member of the petty bourgeoisie in the years after her husband's death and deplored the source of her income: the labor and dues of her peasants. Although the author admitted that Dashkova was a sensible rather than abusive administrator, he argued that in her dealings with her peasants she was severe, collecting as much as seven rubles annually in quit-rent from her serfs.[24] Ogarkov went on to indict Dashkova for failing to adhere to a higher moral standard than her contemporaries, commenting that she was typical of the widowed noblewoman of her era, who often surpassed male proprietors in the efficiency and severity of their administration.[25]

In the Soviet era, ideological constraints limited scholarship on Dashkova to her work in the Academy of Sciences and grudging acknowledgement of her impact on the development of eighteenth-century Russian culture. Even after these restrictions were lifted the troubling inconsistency between Dashkova's espousal of Enlightenment ideals and her position as beneficiary of human bondage reemerged as a persistent theme in post-Soviet Dashkova studies. On the basis of the princess's instructions to her bailiffs and peasants, the historian G. A. Veselaia concluded that Dashkova carefully supervised every dimension of her peasants' lives and did all in her power to facilitate a fair outcome when they brought legal suits against each other. At the same time, Veselaia argued that these documents nonetheless reveal the extent to which Dashkova shared the views of her noble contemporaries, regarding her peasants as no more than property that could be sold or mortgaged at will, and refusing to take advantage of an 1807 decree that permitted proprietors to manumit their serfs with land.[26] By contrast, in a more recent work, Larisa Tychinina dismisses such evaluations as anachronistic, observing that Dashkova's superb administrative skills vastly improved the standard of living of her peasants. Tychinina goes on to argue that to condemn Dashkova's acceptance of serfdom is to ignore the historical context of late eighteenth-century Russia.[27]

For her part, Dashkova consistently presented herself as a benevolent proprietress, whose peasants flourished under her benign rule. Her memoirs abound with recollections of her efforts to improve her estates, which resulted, she claimed, in the affluence of her serfs. Upon the princess's return to Russia from Europe in 1782,

Catherine offered her an estate of 2,500 peasants in Belorussia. Initially, Dashkova rejected the offer, on the grounds that she was not capable of administering "peasants who were half Polish and half Jewish." Although admitting that her administration had rendered her own serfs "more industrious, richer, and happier than they ever had been before," Dashkova despaired of achieving equal success with peasants "of whose customs and language I was quite ignorant."[28] Nonetheless, Dashkova soon yielded and accepted the empress's gift with good grace. Her initial impression of Krugloe was that the peasants there were "filthy, idle, poor, and completely addicted to drink."[29] Six years later, however, Dashkova visited Krugloe and reported with considerable complacency that "my peasants were less wretched and less lazy," and that they "esteemed themselves happier now than they had been under a Polish master or under the administration of the Crown."[30] Similarly, after retiring at the end of her life to her beloved estate near Moscow, Troitskoe, Dashkova found particular consolation in "the prosperity of [her] peasants," whose numbers had increased from 840 to 1,550 during the forty years of her administration. "The number of women . . . had increased even more, since none of them wanted to marry outside my property, while girls of neighboring estates would often marry peasants from Troitskoe," she recounted with satisfaction.[31]

In her memoirs, Dashkova also seized the opportunity to advertise the modest dues she exacted from her peasants, and the warmth and gratitude they displayed in return—a picture considerably at odds with that presented by the princess's detractors. Having taken control in 1794 of the estate of her daughter, Anastasiia, in Kursk province[32] after the latter fell into debt,[33] Dashkova claimed: "The dues I imposed on her peasants were so moderate that they considered themselves lucky indeed."[34] As a result, she added, the revenue of the estate was hardly sufficient to pay off Anastasiia's creditors. The peasants at Korotovo, the estate of her son Pavel in Novgorod province, treated Dashkova with particular respect during her exile there after the death of Catherine II. "We have become rich under your management," they told her, and declared that their quit-rent was lower than that on neighboring estates.[35]

Although Dashkova clearly reveled in her activities as estate manager, she repeatedly emphasized that she labored not for her own benefit, but for the welfare of her dependants. While historians have cast a skeptical eye upon Dashkova's assertions in the memoirs that she held herself fully accountable for the well-being of her serfs, the princess's references to her peasants in her letters leave little doubt that, even if self-interest played a role in her solicitude, her sentiments were sincere.[36] These documents also vividly depict the burdens of absentee landownership in Dashkova's era, as well as the problem of lawlessness in the provinces, which subjected noble and peasant alike to the whims of rapacious neighbors and officials. In a revealing communication to her brother Aleksandr, Dashkova expressed frustration as she recounted her efforts to settle affairs in Krugloe, the estate she had received from Catherine II in 1782. "My health is suffering from my stay here," she lamented. "No pecuniary advantage alone could force me to remain, but when it concerns the welfare of 2,950 beings of both sexes of the same species as me and whom fate has entrusted to my care, I must sacrifice myself." The following day she returned to this theme, deploring the suffering her peasants had endured at the hands of neighboring landlords in her absence. "I reproach myself bitterly that it has been under my administration that they have been martyred in this fashion," she wrote.[37] This incontestable vulnerability of the peasantry, vis-à-vis officialdom and neighboring proprietors, lay at the heart of Dashkova's argument that serfdom was a necessary evil in eighteenth-century Russia and her refusal to acknowledge that the institution should be abolished.

Dashkova, Diderot, and the Defense of Serfdom

The encounter between Dashkova and the French *philosophe* Denis Diderot, and their dispute over serfdom occupy a central place in the memoirs and in scholars' assessments of Dashkova as a representative of the Enlightenment. According to Dashkova, Diderot introduced the topic, criticizing the persistence of "slavery" in the enlightened absolutism of Catherine II. Far from conceding his point, Dashkova responded with a spirited defense, arguing that the nobility acted as indispensable intermediaries between peasants and the state. Dashkova confessed that at one time she had shared Diderot's opinion and "thought of giving more freedom to my peasants and making them happier." She in no way elaborated on the methods she might have employed to augment the liberty of her serfs,[38] but justified her change of heart on the grounds that her peasants "merely became more liable to be pilfered and robbed by every little employee of the Crown."[39]

Significantly, Dashkova drew an explicit connection between the necessity of serfdom and the nature of autocratic government: as long as autocracy and the arbitrary exercise of power survived in Russia, it was the duty of noble proprietors to defend their serfs against corrupt officialdom. This observation was well in keeping with Dashkova's conviction that individual proprietors alone could take sufficient interest in the prosperity of their human property, while representatives of the state sought only to maximize profits. "This is why the most wretched of all peasants are those that belong to the Sovereign," she maintained.[40] After Diderot pointed out that emancipation would eventually augment the fortunes of both noble and peasant, Dashkova again objected, and insisted that the Russian peasantry was ill-prepared to take advantage of freedom. "When the lower classes of my fellow citizens become more enlightened they will deserve to be free," she declared, "because they will know how to enjoy freedom without detriment to their fellows."[41] Dashkova went on to elaborate the factors that mitigated the lot of serfs in her native land: although Russian law forbade serfs to petition the monarch with complaints against their masters, it allowed for provincial officials to relieve abusive proprietors of their peasants and place them in the care of a board of guardians. Thus, the power of serf owners, unlike that of the sovereign, was not unlimited. Dashkova underscored her argument by comparing the emancipated serf to a blind man, unaware of the dangers that surround him, whose sight is restored to him and ends up dying from fear when he realizes the true peril in which he lives. At this, Diderot capitulated, albeit with reluctance. "'What a woman you are,' he burst out," Dashkova recalled with triumph. "'You have upset ideas I cherished and upheld for twenty years.'"[42]

Dashkova's recollection of this episode has the flavor of a set piece. Although Diderot left a record of his meetings with the princess, whom he admired, at no time did he refer to their debate on serfdom, which may not, in fact, have taken place. Moreover, we may safely conclude that Diderot did not revise his negative view of human bondage. Upon his return from Russia in 1784, the philosopher wrote that the only means of preventing the abuses of serfdom was to abolish the institution altogether. At the same time, he conceded that emancipation could not take place as long as those who controlled human property refused to acknowledge its evils, and while serfs themselves remained unaware of the benefits of freedom.[43] With this admission, Diderot betrayed the ambivalence of so many Enlightenment thinkers, who advocated the rights of man and deplored the debasement of the greater part of the population, yet simultaneously perceived the "lower classes" as ignorant, backward, and incapable of governing themselves.

The Dilemma of Emancipation

Dashkova's support for the institution of serfdom troubled early historians and continues to perplex modern commentators. Certainly, if—as Diderot observed in a letter to the princess—liberty was the fundamental characteristic of their age,[44] Dashkova's credibility as a disseminator of enlightenment values among her compatriots appears highly suspect. Until the end of her life, the princess expressed steadfast support for advancing the freedom of the nobility, while rejecting the liberation of Russia's peasant population as unviable. In light of the fragility of noble property rights and corporate identity in Russia, however, Dashkova's defense of unfree labor was consistent with her promotion of noble prerogatives and her related preoccupation with limiting the power of the monarch. Nor were her views completely at odds with those of European intellectuals: the French *philosophes* themselves displayed a range of opinion on the topic of emancipation and its implementation. Voltaire argued in favor of doing away with serfdom, but against granting land to the peasantry, while Rousseau advocated a form of gradual emancipation.[45] Even the most radical proponent of the abolition of serfdom in eighteenth-century Russia, Aleksandr Radishchev, favored emancipation over time, rather than the immediate creation of a free peasantry. Upon reading an essay in which Radishchev eulogized a fellow Russian critic of serfdom, Dashkova remarked that "his thoughts or expressions were dangerous in times in which we lived"[46] and feared that his writings would put him in danger—a prophecy that was fulfilled when Radishchev was exiled to Siberia after the publication in 1790 of his abolitionist tract, *A Journey from St. Petersburg to Moscow*. For Dashkova to have championed the abolition of serfdom would have made her a remarkable figure indeed among her contemporaries.[47]

The apparent incongruity of Dashkova's advocacy of serfdom was the inevitable outcome of the tension in her dual roles as the leading representative of the Enlightenment in Russia and a tireless advocate of noble interests. Indeed, Dashkova's aspiration to secure the liberty of the nobility vis-à-vis the sovereign, along with her awareness of the enduring fragility of noble status, appears not only in the memoirs, but also in her other literary artifacts. Dashkova's criticism of despotic government was in part manifested by her admiration of the English system of constitutional monarchy. In an account written after her first journey to England in 1770, titled "The Journey of a Russian Noblewoman through Certain English Provinces," Dashkova remarked that "England was more to my liking" than any other nation in which she had traveled. While commenting favorably on English agriculture, education, and manners, Dashkova also observed that the English form of government also surpassed that of other nations.[48] Elsewhere, in her account of her sojourn in Italy in the memoirs, she devoted an extensive passage to the constitution of the city of Lucca, which she admired for its aristocratic and elective nature, and for its separation of powers.[49]

Dashkova's espousal of enlightened absolutism as the best form of government for Russia was analogous to her views on serfdom: for all her admiration of liberty, she never wavered from her conviction that Russian society could not advance without the guidance of a strong monarch, albeit one subject to the rule of law. In this regard, her views resembled those of her compatriots.[50] "Limited monarchy, headed by a master who could be a father to his people . . . must be the goal of every thinking person," Dashkova declared. "For power, lodged with the masses, whose opinions are lightly held and forever changing, is unstable."[51] Dashkova's esteem for limited monarchy long predated her writing of the memoirs: after her meeting with Diderot, thirty years before the princess composed her celebrated work, the *philosophe* remarked upon Dashkova's aversion for all forms tyranny and despotism.[52] Yet distaste for arbitrary rule and admiration for constitutional government by no means prompted either Dashkova or her contemporaries to advocate abolishing the monarchy or to underestimate the vital role the sovereign played in the development of Russian society.[53]

For all her zeal to elevate the status of the nobility, Dashkova makes no mention in her memoirs of the 1762 Manifesto liberating Russian noblemen from obligatory service to the state, nor of Catherine II's 1785 Charter to the Nobility, which, for the first time, articulated the rights of the nobility in written law. Nonetheless, Dashkova's attention to the significance—and fragility—of noble prerogatives in Russia surfaces repeatedly in the memoirs. Among her criticisms of Peter the Great, Dashkova condemned the emperor for undermining the rights of the gentry and the rule of law. Despite his genius, she argued,

"[H]e treated all people without distinction as slaves who must bear with everything . . . He would not have weakened the power of the law and the respect which is due to it if he had not changed laws so often, including his own."

Tellingly, Dashkova then turned to the subject of Peter's behavior toward the nobility. "He almost entirely abolished the freedom and privileges of the gentry as well as of the servants," she observed. Dashkova also pointed out that when Peter built St. Petersburg, his window on the West, "thousands of workmen perished in that marsh, while he thrust upon the nobles the burden of finding and supplying more workmen, and forced them to build brick houses for themselves in accordance with *his* plans, regardless of whether they needed houses in Petersburg or whether they did not."[54] When applying for permission to leave Russia in 1768, which she required as lady-in-waiting to the empress, Dashkova noted that "as a member of the gentry class I had full right to [travel abroad] without asking permission first"[55]—a privilege that her noble predecessors had not enjoyed under previous monarchs. Finally, in a chronology compiled in 1807, Dashkova singled out manifestos granting noble proprietors the right to profit from natural resources on their estates and the 1785 Charter to the Nobility as events of special significance in the history of Russia.[56]

Ultimately, the princess's ruminations on serfdom occupy a lesser place in the memoirs than a far more urgent theme: the consequences of unchecked and arbitrary power on the part of the sovereign. Following Catherine II's death in 1796, Dashkova not only witnessed the attempts of Catherine's son and successor, Paul, to destroy his mother's work and revoke the privileges of the nobility, but was herself the victim of persecution when she was exiled to her son's remote village, Korotovo. Although her own banishment endured only a year, Dashkova remarked more than once in her memoirs on the deportations and arrests that followed Paul's accession to the throne and on the atmosphere of "general terror," observing "there was not a single family among the nobility that did not have at least one member either exiled in Siberia or confined in a fortress."[57] To extend liberty to the peasantry, when the nobility itself had failed to achieve genuine political freedom or security with relation to the sovereign, was unthinkable.

Conclusion

In light of her status as a prominent *femme savante* and political figure, Princess Dashkova's defense of serfdom was only one of many characteristics that perplexed both her admirers and detractors. Time and again, Dashkova depicted herself as a beneficent proprietress, whose vocation in life was not political advancement, but the care of her dependants: her children and her serfs. In the years following Catherine II's death and Dashkova's expulsion from court, the latter claimed that while she lived in retirement, she was "solely occupied with the welfare of my peasants."[58] She was also keenly aware of the potential for abuse in the institution of serfdom.[59] At the same time, the memoirs leave no doubt that Dashkova regarded her peasants as property, to be treated with benevolence but whose role was to do her bidding,[60] and her reaction to peasant disorder can well be imagined. Even when contemplating her own demise, the princess never wavered from her conviction that serfdom was a necessary evil in her era: the terms of her will allowed for the manumission of only two of her female house serfs, in return for years of devoted service.[61] Yet Dashkova was far from an anomaly. Her failure to advocate the abolition of serfdom was emblematic of the tension in Enlightenment philosophy between the celebration of freedom and the reality of governing an imperfect society. In this regard, as in so many others, Dashkova epitomizd the incongruities of her era.

Notes

1. Peter Gay, *The Enlightenment: An Interpretation. Vol. 2: The Science of Freedom* (New York, 1969), pp. 407–23.

2. Bernard Bailyn, *To Begin the World Anew: The Genius and Ambiguities of the American Founders* (New York, 2003), pp. 40, 48–49.

3. Isabel de Madariaga, *Russia in the Age of Catherine the Great* (New Haven, 1981), pp. 132–36.

4. On the central role of serfdom in noble life, see Priscilla R. Roosevelt, *Life on the Russian Country Estate: A Social and Cultural History* (New Haven, 1995).

5. D. I. Ilovaysky (1884); quoted in Robert Argent Longmire, *Princess Dashkova and the Intellectual Life of Eighteenth Century Russia* (Master's thesis, University of London, 1955), p. 316.

6. Longmire, *Princess Dashkova and the Intellectual Life*, p. 72.

7. *Polnoe sobranie zakonov Rossiiskoi imperii*, 1st series (St. Petersburg, 1830), vol. 22, no. 16.187 (April 21, 1785), sect. A, art. 6.

8. Among wealthy aristocratic families in Russia, it was common practice to designate a sum of money "for the purchase of estates" in dowry contracts, rather than to fragment the family estate by bestowing villages upon the bride. For an English translation of Dashkova's dowry, see Robin Bisha, Jehanne M. Gheith, et al., *Russian Women, 1698–1917: Experience and Expression. An Anthology of Sources* (Bloomington, 2002), pp. 60–63. On women's inheritance rights and the dowry in imperial Russia, see Michelle Lamarche Marrese, *A Woman's Kingdom: Noblewomen and the Control of Property in Russia, 1700–1861* (Ithaca, 2002), pp. 17–43, 138–41. The terms of Count Vorontsov's will excluded Dashkova from any further claims on his estate: Rossiiskii gosudarstvennyi arkhiv drevnikh aktov (RGADA) (Russian State Archive of Ancient Acts), f. 1261 (Vorontsovy), op. 1, ed. khr. 8, ll. 1–3.

9. E. R. Dashkova, *The Memoirs of Princess Dashkova*, trans. and ed. by Kyril Fitzlyon (Durham, 1995), pp. 64–65. Dashkova describes this land in her memoirs. The fate of her dowry appears in her will, composed in 1807, and in petitions to Catherine II, in which Dashkova explained why she should inherit more than the one-seventh of her husband's estate which she should have received according to Russian inheritance law. See RGADA, f. 1261, op. 1, ed. khr. 30, ll. 4, 8–10 and Rossiiskii gosudarstvennyi istoricheskii arkhiv (RGIA) (Russian State Historical Archive), f. 1151 (Departament grazhdanskikh i dukhovnykh del Gosudarstvennogo soveta), op. 1, ed. khr. 8, l. 2.

10. Dashkova, *Memoirs*, p. 64.

11. *Materialy dlia biografii kniagini E. R. Dashkovoi* (Leipzig, 1876), pp. 122–25. This undated letter exists, although not in the original, in manuscript form in the British Library: Additional Manuscript no. 31, 911 (Hardwicke Papers), ff. 245–47.

12. Dashkova, *Memoirs*, p. 149.

13. Dashkova, *Memoirs*, p. 114.

14. Women were far less likely than men to receive estates from the monarch, despite the allegations of

nineteenth-century historians that during the reign of empresses noblewomen were frequently the recipients of settled estates; see Marrese, *A Woman's Kingdom*, pp. 137–38. Many ladies-in-waiting received dowries from the empress, which could range from fairly modest personal effects to several thousand rubles. See RGADA, f. 14 (Pridvornoe vedomstvo), op. 1, ed. khr. 98, for an inventory of dowries bestowed during the reign of Empress Elizabeth (1747–1762). In return for Dashkova's support in 1762 and, later, for her work at the Academy of Sciences, Catherine eventually awarded the princess sums that amounted to well over fifty thousand rubles; see Dashkova, *Memoirs*, p. 88; L.V. Tychinina, *Velikaia rossiianka: zhizn' i deiatel'nost' kniagini Ekateriny Romanovny Dashkovoi* (Moscow, 2002), p. 120; RGIA, f. 468 (Kabinet ee Imperatorskogo Velichestva), op. 1, ed. khr. 3886, l. 11; ed. khr. 3897, l. 172.

15. This discussion of noblewomen's property rights is drawn from Marrese, *A Woman's Kingdom*.

16. Larry Wolff, *Inventing Eastern Europe: The Map of Civilization on the Mind of the Enlightenment* (Stanford, 1994).

17. As early as 1719, Peter I decreed that the estates of serf owners who mistreated their peasants should be placed under guardianship. During the eighteenth century, this provision was largely ignored and abusive proprietors were punished only in the most extreme cases. In the first half of the nineteenth century, however, state representatives developed a veritable obsession with preventing the mistreatment of serfs, in the hope of minimizing the possibility of peasant revolt. See Jerome Blum, *Lord and Peasant in Russia from the Ninth to the Nineteenth Century* (Princeton, 1961), pp. 435–39 and Marrese, *A Woman's Kingdom*, pp. 229–37.

18. Quoted in Peter Kolchin, *Unfree Labor: American Slavery and Russian Serfdom* (Cambridge, 1987), p. 42. Kolchin's work remains the classic study on slavery and serfdom.

19. RGADA, f. 1261, op. 1, ed. khr. 30, ll. 4, 8–10.

20. Dashkova, *Memoirs*, p. 99.

21. See Marrese, *A Woman's Kingdom*, pp. 171–204.

22. Martha Wilmot Bradford, *The Russian Journals of Martha and Catherine Wilmot, 1803–1808*, eds. Marchioness of Londonderry and H. M. Hyde (London, 1935), p. 201.

23. Martha Wilmot Bradford, *The Russian Journals*, pp. 223–24.

24. In her will, Dashkova claimed that she collected only three rubles a year in *obrok* (quit-rent) from her serfs on the estates she inherited from her husband and, as a result of her "mild administration," the number of peasants had increased substantially. RGADA, f. 1261, op. 1, ed. khr. 30, l. 12.

25. V.V. Ogarkov, *E. R. Dashkova: ee zhizn' i obshchestvennaia deiatel'nost'* (St. Petersburg, 1893), pp. 36–39. See also V. I. Semevskii, who accused Dashkova of being an "extremely severe" proprietress: "Kniagina Ekaterina Romanovna Dashkova," *Russkaia starina*, vol. 9 (1874), pp. 417–18.

26. G. A. Veselaia, "Ekaterina Romanovna Dashkova v sele Troitskom (Materialy k biografii)," *Materialy po istorii russkoi kul'tury kontsa XVIII-pervoi poloviny XIX veka. (Trudy gosudarstvennogo ordena Lenina istoricheskogo muzeia)(Vypusk no. 58)* (Moscow, 1984), pp. 81–82; "Poslaniia kniagine E. R. Dashkovoi svoim krest'ianam v Novgorodskuiu guberniiu v derevniu Korotovo," in *E. R. Dashkova i ee vremia: issledovaniia i materially*, eds. L.V. Tychinina et al. (Moscow, 1999), pp. 113–24. See also L. Ia. Lozinskaia, *Vo glave dvukh akademii*. 2nd ed. (Moscow, 1983), p. 71 and A. Nivier, "E. R. Dashkova i frantsuzskie filosofy Prosveshcheniia Vol'ter i Didro," in *Ekaterina Romanovna Dashkova: issledovaniia i materially*, eds. A. I. Vorontsov-Dashkov et al. (St. Petersburg, 1996), pp. 47–48.

27. Tychinina, *Velikaia rossiianka*, p. 176.

28. Dashkova, *Memoirs*, p. 194.

29. Dashkova, *Memoirs*, p. 195.

30. Dashkova, *Memoirs*, p. 227.

31. Dashkova, *Memoirs*, p. 248.

32. Dashkova's daughter, Anastasiia Shcherbinina, eventually regained control of the estate, which consisted of almost one thousand serfs: however, she did not share her mother's administrative talents. In her will, Shcherbinina left instructions for her cousin, Mikhail Semenovich Vorontsov, to sell the estate, as it was clearly in debt, and distribute the proceeds from the sale among her three wards, the illegitimate children of her deceased brother, Pavel Dashkov (RGADA, f. 1261, op. 1, ed. khr. 55, ll. 1–5).

33. RGADA, f. 1239 (Dvortsovyi otdel), op. 3, ch. 80, ed. khr. 40358.

34. Dashkova, *Memoirs*, pp. 242–43.

35. Dashkova, *Memoirs*, p. 269.

36. Dashkova acknowledged that mistreatment and neglect would serve only to diminish the income her peasants generated on her estates. See E. N. Marasinova, *Psikhologiia elity rossiiskogo dvorianstva poslednei treti XVIII veka (Po materialam perepiski)* (Moscow, 1999), pp. 207, 210. During her exile in 1797, Dashkova petitioned Emperor Paul to allow her to spend several weeks a year in Moscow, in order to oversee the management of her serfs and to prevent the diminution of her income. At this time, permission was refused (RGADA, f. 1239, op. 3, ch. 111, ed. khr. 58231).

37. *Arkhiv kniazia Vorontsova* (Moscow, 1872), vol. 5, pp. 191, 193.

38. Dashkova's options for expanding the freedom of her peasants were, in fact, limited in the eighteenth century. Although serf proprietors could free individual serfs, estate owners could not legally manumit entire villages. Manumitted serfs became members of the state peasantry.

39. Dashkova, *Memoirs*, p. 124.

40. Dashkova, *Memoirs*, p. 195.

41. Dashkova, *Memoirs*, p. 125.

42. Dashkova, *Memoirs*, p. 125.

43. Quoted in Lozinskaia, *Vo glave dvukh akademii*, pp. 49–50. Diderot consistently referred to serfdom as slavery in his writings.

44. Denis Diderot, *Correspondance*, ed. Georges Roth (Paris, 1964), vol. 11, p. 20.

45. E. Haumant, *La culture française en Russie* (Paris, 1913), p. 140.

46. Dashkova, *Memoirs*, p. 236.

47. Dashkova's views were the rule, rather than the exception, among educated Russians. See G. R. Derzhavin, *Sochineniia Derzhavina*. 2nd ed. (St. Petersburg, 1876), vol. 6, p. 778. In her memoirs, Catherine II noted that fewer than twenty people at the Russian court shared her views in favor of serf emancipation—a point of view she would abandon over the course of her reign. Quoted in Rodger Bartlett, "Poselenie inostrantsev v Rossii pri Ekaterine II i proekty osvobozhdeniia krepostnykh krestian," in *Evropeiskoe Prosveshcheniie i tsivilizatsii Rossii*, ed. S. Ia. Karp (Moscow, 2004), 261. As Isabel de Madariaga observes, however, during the debates in the Legislative Commission, a committee of high nobles argued in favor of the right of proprietors to manumit not only individual serfs, but entire villages, and for the creation of "free estates," in which peasants would be contractually bound to the soil, but personally free. See Madariaga, *Russia in the Age of Catherine the Great*, pp. 174–78.

48. "Puteshestvie odnoi rossiiskoi znatnoi gospozhi po nekotorym angliiskim provintsiiam," in E. R. Dashkova, *O smysle slova "vospitanie": sochineniia, pis'ma, dokumenty* (St. Petersburg, 2001), p. 95.

49. Dashkova, *Memoirs*, pp. 167–69.

50. On educated society's support in eighteenth-century Russia for the monarchy, see Cynthia Hyla Whittaker, *Russian Monarchy: Eighteenth-Century Rulers and Writers in Political Dialogue* (DeKalb, Illinois, 2003).

51. Dashkova, *Memoirs*, p. 60.

52. Diderot, *Correspondance*, vol. 10, p. 167.

53. Marasinova, *Psikhologiia elity*, pp. 134–57.

54. Dashkova, *Memoirs*, p. 181.

55. Dashkova, *Memoirs*, p. 117.

56. British Museum, Additional Manuscript no. 31, 911, ff. 212–13.

57. Dashkova, *Memoirs*, p. 250.

58. E. R. Dashkova, *Memoirs of the Princess Daschkaw, Lady of Honour to Catherine II*, ed. Mrs. W. Bradford (London, 1840), vol. 1, p. xxxvii.

59. In an article titled "Portraits of My Family," published in the journal *Sobesednik* in 1784, Dashkova drew particular attention to the beatings suffered by house serfs in the family she describes. See "Kartiny moei rodni, ili Proshedshie sviatki," in Dashkova, *O smysle slova "vospitanie,"* pp. 151–53.

60. See Dashkova's description of the entertainment she organized for Mrs. Hamilton when the latter visited

Troitskoe. "I had all the peasants…gathered together, all dressed in their best clothes," she wrote. "The weather was gorgeous, and I made them dance on the grass and sing Russian songs" (Dashkova, *Memoirs*, p. 218).

61. Nowhere in the memoirs does Dashkova mention the peasant uprising of 1773–1775, led by Emel'ian Pugachev, which devastated the provinces and in which several thousand nobles and officials were murdered. She does refer, however, to peasant revolts that took place soon after Paul's ascension to the throne, but claimed that her serfs refused to engage in any kind of disorder. Her peasants, Dashkova declared, "said that they felt themselves more fortunate than the peasants who belonged to the Crown" (Dashkova, *Memoirs*, p. 263). For Dashkova's will, see RGADA, f. 1261, op. 1, ed. khr. 30, l. 15.

Virtue Must Advertise:

Self-Presentation in Princess Dashkova's Memoirs

MARCUS C. LEVITT

*P*rincess Ekaterina Romanovna Dashkova was one of the most colorful and striking figures of the age of Catherine the Great, itself an epoch of oversize personalities. "Catherine the Little," as Dashkova refers to herself in her memoirs, was—next to the empress Catherine—the most prominent and commented-upon Russian woman of her day. Political activist, author, editor, courtier, first woman to head the Imperial Academy of Sciences and Arts in St. Petersburg and founder of the Imperial Academy of the Russian Language, Dashkova was arguably also Russia's first modern female celebrity. Much like Benjamin Franklin, she captured the imagination of the educated world as her country's de facto cultural ambassador. While Franklin personified homegrown

American democracy, Dashkova was emissary for Russia's special brand of "Enlightened absolutism." Dashkova's main claim to fame was not merely as the extraordinary Russian woman intellectual, but also, and more notoriously, as the principal co-conspirator, at the tender age of nineteen, in the 1762 palace "revolution" that raised Catherine to the throne and deposed and dispatched Peter III.[1]

A Triple Defense

In contrast to Franklin's *Autobiography*—a collection of materials begun mid-career as a family history recounted for his son and filled with stories of a life rich in success and public recognition—Dashkova's memoir was written near the end of her life as an attempt to rescue her public image from oblivion or worse, misrepresentation. Hers is a purposeful *apologia pro vita sua* presented to the court of posterity and public opinion. While Franklin took up the pen at a time when he was approaching the zenith of international popularity, Dashkova's memoirs were written when her celebrity was in almost total eclipse. Her matter-of-fact and generally unsmiling tone also stands out against Franklin's calm and lightly self-deprecating irony; it reflects her more serious purpose of defending her life's legacy and that of Catherine the Great's Russia.

That legacy was called into question almost immediately after Catherine's death in 1796, an event Dashkova described in her memoirs as "a blow . . . which for Russia represented the greatest possible disaster."[2] Catherine's son, the new emperor Paul I, undertook a campaign to rehabilitate the honor of his ignominiously deposed father, stripping Dashkova of her official positions and sending her into a brief exile in northern Russia. Even after her return to imperial grace following the accession of Alexander I in 1801, Dashkova was indignant to find "the people surrounding the Emperor . . . unanimous in disparaging the reign of Catherine II and in instilling in the young monarch the idea that a woman could never govern an Empire."[3] This misogynist attitude, which had been codified under Paul in a new law of succession based on male primogeniture, also clearly cast its shadow on Dashkova, "one of the first women in Europe to hold governmental office."[4]

Perhaps just as importantly, Dashkova took up her pen to refute what she referred to as the "flood of pamphlets libeling Catherine II."[5] The eighteenth century had seen the flowering of a category of literary work that came to be known as "Russica" (writing about Russia) that debated whether or not the country should be considered part of European Enlightenment civilization.[6] Catherine's detractors, motivated by long-standing political animus, shoveled dirt on the empress's personal life and her court. Dashkova was outraged by the "cleverly concocted lies and foul fictions" spread by "certain French writers," who "at the same time undertook to blacken and slander her innocent friend"; that is, Daskova herself.[7] Any stains on the empress's "spotless reputation" threatened to tarnish Dashkova's own.

Dashkova's memoir is thus a triple defense: it is a vindication of Catherine the Great as a truly "great and enlightened empress;"[8] an affirmation of Russian Enlightenment culture; and, not the least, a justification and clarification of Dashkova's own historical role. The memoir spans Dashova's entire life up through the time of its writing. It focuses on the story of the "revolution" that brought Catherine to the throne; Dashkova's two extended trips to Europe; her public life in St. Petersburg; her exile under Paul I; and, briefly, her last years. The original French title of the work, *Mon histoire,* which may mean both "My Story" and "My History," and suggests the merging of an individual and historical narrative. While historians dispute aspects of Dashkova's account, most notably, the assertion that she played the key role in the conspiracy that brought Catherine to power, the very weaknesses in historical accuracy or memory that may exist in *Mon histoire* turn it into an all the more eloquent exercise in constructing a self-image. In the remainder of this essay, I will attempt to describe that image, placing it within the context of the cultural values that Dashkova champions in her memoirs.

Making Virtue Visible

We may begin to seek the roots of Dashkova's strikingly powerful sense of self in the circumstances of her early life as she describes them. On the one hand, Dashkova recounts her overwhelming desire for approbation and love, and on the other, tells of her desperate sense of being lonely and "wounded by indifference"—states of mind sparked by the loss of her mother at age two. "I became serious-minded and studious . . . Reading soothed me and made me happy. . . ." The life of the mind and the satisfactions of superior intellect offered a compensation, as she resolved to become "all I could be by my own effort . . . [in a] presumptuous effort to be self-sufficient."[9] She was attracted in particular to Enlightenment political and educational theory, an arena in which her "relentless curiosity" could be satisfied. Her memoirs offer a sophisticated defense of Enlightened selfhood, as Dashkova constructs and defends a powerful, charismatic, intellectually impressive image of an ideal public self. Struggling to describe the "peculiarities & inextricable varietys" of Dashkova's contradictory character, her friend Catherine Wilmot concluded:

> For my part I think she would most be in her element at the Helm
> of the State, of Generalissimo of the Army, or Farmer General of
> the Empire. In fact she was born for business on a large scale which
> is not irreconcilable with the Life of a Woman who at 18 headed a
> Revolution & who for 12 years afterwards govern'd an Academy of
> Arts & Sciences.[10]

Dashkova fully subscribed to the ideal of her age, defining herself in terms of the "Great Man." As the label indicates, the serious public role she chose to emulate was culturally gendered as male; indeed, the classical Roman heritage that was one of its main sources preserved a direct etymological linkage between the male (*vir*) and virtue (*virtus*) itself.[11]

At the same time, Dashkova's memoir—by the author's no less eloquent testimony—reveals the dark and unhappy underside of the "Great Man." Dashkova was by her own admission plagued by constant physical ailments, as well as by a "deep dejection" and "bitter disappointments" that haunted her existence, which we might conceptualize as her frustrated "female" shadow self-demanding its due. Dashkova senses that the protective façade of superior intellect is "presumptuous," and from the very beginning of her conscious life fears that "my sensibility and weak nerves would ruin my life by making it impossible to bear the pain of disappointment and wounded pride . . . I was beginning to have the foreboding that I would not be happy in this world."[12] Dashkova's need for praise, so poignant for a motherless child, was also, and perhaps even more importantly, an especially powerful directive of her age; that was, *the need to be seen and approved*.[13] The need for approval is undoubtedly a universal human necessity, yet self-display—whether in court ceremony, on the stage, in architecture, urban planning, landscape gardening, clothing, or the fine arts in general—took on special prominence as a cultural imperative during Russia's early modern age. It offered visible proof of Russia's imperial grandeur, demanding recognition of national greatness to vie with that of the West. Beyond its usual function as a simple marker of power and prestige, visual display also played a key role in Russian Enlightenment thought and self-consciousness: to put it baldly, virtue must advertise.[14] Visibility, and chiefly the visibility of virtue, became a particularly important issue in Catherine the Great's political program, especially in the early part of her reign. Catherine justified her assumption of power by means of her superior Enlightenment credentials: she who was self-evidently best qualified to rule, and most virtuous in promoting the public welfare, deserved to rule.[15] Dashkova, having indissolubly linked her fortune with Catherine, fully ascribed to this political and moral program.

Virtue Under Siege

A useful episode for understanding the drama of Dashkova's self-presentation is her description of the crisis that followed Catherine's death. The new emperor Paul ordered her to leave Moscow and retire to her place in the country, where she was instructed to "ponder on the events of the year 1762" and to await his decision on her further fate. Dashkova writes:

> I left Moscow on 6 December. My health was reduced to a struggle against death. Every other day I wrote to my brother and other members of my family, who also wrote very regularly to me. Several of them, including my brother, told me that Paul I's behavior toward me was dictated by what he thought he owed to his father's memory, but that at his coronation he would change our fate. I shall quote my reply to my brother as one of the many prophecies I have made which have come true:

"You tell me, friend, that after his coronation Paul will leave me alone. You do not know him then. Once a tyrant begins to strike he continues to strike until his victim is totally destroyed. I am expecting persecution to continue unabated, and I resign myself to it in the full submission of a creature to its Creator. The conviction of my own innocence and lack of any bitterness or indignation at his treatment of me personally will, I trust, serve me in place of courage. Come what may, and provided he is not actively malevolent to you and those near and dear to me, I shall do or say nothing that will lower me in my own eyes. Goodbye, my friend, my well-beloved brother. All my love." [16]

Here Dashkova stands tall on the stage of history. She presents herself like a heroine of tragic drama or a sentimental novel, a "Great Man" displaying the transparency of her virtue for all to appreciate.[17] She describes innocence and virtue pitted against relentless malice, virtue under physical and emotional siege. Her response brings her courage and self-possession into sharp relief. The quoting of the letter both helps Dashkova establish the documentary nature of the moment—its historical truth—and reflects her exalted, extremely "literary," self-image; it is as if she were reciting a tragic monologue or contemplating herself at a remove, as in a mirror.

While Dashkova's dramatic stance might seem appropriate considering the real threat from Paul, similar extreme oppositions—a continual struggle between life and death, salvation and destruction, approbation and opprobrium—operate throughout the text. They characterize Dashkova's understanding of the self as in a constant struggle between absolute virtue and vice, whose outcome has highly serious, even metaphysical, consequences. Dashkova presents herself as totally virtuous, and she makes no secret of the pride and self-satisfaction she feels in her virtue. Although her "submission" to God's will may have something in common with the Russian Orthodox notion of kenosis (the "emptying of the self" in imitation of Christ), Dashkova's language stresses more her adherence to an Enlightenment conviction—strongly echoing classical Stoicism—of righteousness founded on reason and superior self-knowledge. This submission is not humility in the traditional religious sense, born of a sense of sinfulness or guilt, but a defense of pride and self-esteem as an enduring virtue.

The Denial of the Personal

What may seem strange, particularly to a modern sensibility, is the extent to which Dashkova and the people of her epoch equated the (good) self with universal and "natural" merit. As in classicist tragedy, the personal or private element, if not in harmony with the demands of family, society, and Nature, is ascribed to the dark side. Conversely, the virtuous self is in perfect accord with the absolute and universal. For Dashkova, to be virtuous is to act unselfishly, "disinterestedly"; conversely, to act in the name of "personal interest" is evil. "Private" merit can only consist of impersonal virtue, and to act in one's self-interest selfishly is to act in an evil way. Altruistic self-sacrifice is the measure of goodness. The following passage, in which Dashkova describes the consolations of pride in the face of suffering, is characteristic:

> I never pursued either my personal interests or the criminal elevation of my own family . . .

> I . . . gathered support from the feeling of my own innocence, the purity of my conscience, and a certain moral pride which gave me strength and courage, but which I had never previously suspected in myself and which, after giving the matter much thought, I could only attribute to resignation, a sentiment proper to every rational being.[18]

Dashkova's "resignation" includes a big dose of self-satisfaction, as she elevates herself to the ranks of "rational beings."

Dashkova's English friends Martha and Catherine Wilmot, who convinced Dashkova to write her memoirs, and helped with their actual production, also left several penetrating descriptions of this aspect of her self-image. In speaking of her conspicuous vanity, Martha wrote in a letter to her father:

> [Dashkova's] establish'd opinion of herself is such that, if I can make you feel what I mean, it is as if she was distinct from herself and look'd at her own acts and deeds and character with a degree of admiration that she never attempts to express the expression of, and that with a sort of artlessness that makes one almost forgive her. Her principles are noble and possess'd of influence which extends to *absolute* dominion over the happiness . . . [of] some thousands of Subjects. She invariably exerts it for their welfare . . . As a relation she is everything to her family. . . .[19]

Dashkova presumes "a degree of admiration" for herself that is beyond expression—a conviction so absolute that it suggests her being seen "distinct from herself." What, according to Martha, (almost!) keeps this exalted sense of self-worth from being repellent is Dashkova's "artlessness," her presumption that image and reality match, the firm conviction that "her principles are noble" and disinterested. Martha shrewdly associates this attitude with Dashkova's power, both as a landowner (her sense of "*absolute* dominion" over her serfs [Martha's italics]) and the great influence she exercises over her extended family. Martha senses a clear correlation between Dashkova's assertion of political power and its justification as something "invariably" exerted for the "welfare" of those under her dominion.[20]

In an autocratic context, Dashkova's claim on virtue may be seen as staking a claim on political power. Scholar M. M. Safonov describes her predicament: "Dashkova had the courage to be a personality . . . at a time when only one person in this autocratic country had the right to be a personality—Catherine II."[21] This is the Dashkova who heroically challenged tyrants and who stood up for enlightened ideals. By contrast, as Martha sensed, Dashkova's uncompromising insistence on her own moral authority itself reflected an uncomfortably authoritarian claim on virtue, which, as we have suggested, stemmed from her exalted altruistic conception of the virtuous self.[22]

Dashkova and Franklin: The Right to be an Oddity

Franklin's *Autobiography* offers both some striking points in common with Dashkova's memoirs as well as some sharp contrasts that help clarify the problem of virtue and making it public. Franklin shared with Dashkova a lifelong preoccupation with living a virtuous life. Both he and Dashkova put virtue at the center of their ideal of the good life, and both framed the issue of being virtuous in terms of the good of society. Like Dashkova, Franklin argued that virtue is not of value merely or primarily for is own sake but as the single path to practical well-being. As Franklin put it, "vicious actions are not hurtful because they are forbidden, but forbidden because they are hurtful, the nature of man alone considered."[23] No less than Dashkova, Franklin set very high moral standards, as exemplified in the well-known scheme for self-improvement that he laid out in the *Autobiography*. He set forth to train himself in a list of thirteen leading virtues, an undertaking he described as "a bold and arduous project of arriving at moral perfection."[24] He found implementation even more arduous than originally imagined. When it came to the last virtue on his list, humility, Franklin admitted that "no one of our natural passions [is] so hard to subdue as *pride*." He wrote of its stubborn and paradoxical nature:

> Disguise it, struggle with it, beat it down, stifle it, mortify it as much as one pleases, it is still alive, and will every now and then peep out and show itself; you will see it, perhaps, often in this history; for, even if I could conceive that I had compleatly overcome it, I should probably be proud of my humility.[25]

Like Dashkova, and many other thinkers of the day, Franklin recognized the ambiguous status of pride (vanity, ambition, the desire for approbation) as a natural impulse that may be directed either to the good or the bad. He too argues in defense of what we may call good pride, that which produces "good to the possessor, and to others that are within his sphere of action."[26] At the same time—and unlike Dashkova—he gently ridicules his own autobiographical project as not only offering the model of a life "fit to be imitated" but also as the comforting indulgence of an old man's weakness. Dashkova never admits such weakness, nor does she admit the blemish of "bad pride" in herself. Conversely, Franklin recognizes both the ideal of the virtuous self and the intractable, all-too-human problems of its realization. In coming to grips with his "bold and arduous project of arriving at moral perfection," he notes:

> something that pretended to be reason . . . was every now and then suggesting to me that such extream nicety as I exacted of myself might be a kind of foppery in morals, which, if it were known, would make me ridiculous; that a perfect character might be attended with the inconvenience of being envied and hated; and that a benevolent man should allow a few faults in himself, to keep his friends in countenance.[27]

While the "something that pretended to be reason" might gainsay the pursuit of moral perfection, Franklin nevertheless recognizes the drawbacks of the kind of militant "virtue on display" that Dashkova demands of herself. (And indeed she is constantly on guard against those who ridicule, envy, and hate her.) Where Dashkova insists on strict construction of virtue, and on the complete parity of the inner and outer self, Franklin allows for a degree of dissimulation. He either keeps his "extream nicety as I exacted of myself" *to* himself or, as he notes with regard to the attempt at exercising humility, allows himself the *appearance* rather than the reality.[28]

For Franklin, as for Dashkova, virtue and its recognition were an essentially social phenomena, forged in the crucible of sociability—the self as necessarily mirrored and negotiated through one's peers. At the same time, we need to keep in mind the significant differences between the social and political contexts in which this sociability operated. Franklin describes the world of opinion-makers in colonial Philadelphia, a world in which he was a major player. Dashkova, on the other hand, struggled to assert herself within the restricted and highly stratified setting of the court and European high society. The "absolutist" context of old regime Russia also left its mark on her thinking. On the level of moral theory, Dashkova was a strict constructionist with regard to virtue, and her journalistic writings promote the virtuous life as a necessary goal. For example, she wrote: "Many consider virtue to be harsh and intolerant to human weaknesses (*strogoiu i k chelovecheskim slabostiam nesniskhoditelnoiu*). True, for people suffering from vice, virtue is insupportable, and they can therefore never be happy; but for those who are able to think and feel, nothing is as pleasant as virtue."[29] In sharp contrast to the Pennsylvania democrat, Dashkova was known for her sharp outspokenness, and in her memoirs she often remarks on her inability to restrain and conceal her emotions: "Nature had not endowed me with the gift of pretence, so essential when dealing with Sovereigns and even more with the people round them. Disgust, contempt, indignation—there they all were, writ large on my countenance whenever I felt them."[30] This was more than a simple "natural" lack of self-restraint.

Dashkova, like a heroine in classicist tragedy, refuses to compromise on the truth and insists on speaking her mind, especially in situations involving people in power where such frankness could entail unpleasant consequences.

Martha Wilmot also remarked on the impulsiveness that was a hallmark of Daskova's behavior in society. She noted in her journal:

> It never enters into her head or heart to disguise any sentiment or impulse . . . , & therefore you may guess what a privileged sort of Mortal she makes herself! The Truth is sure to come out whether agreeable or disagreeable, & lucky it is she has sensibility & gentleness of Nature, for if she had not she would be a Public Scourge! She is the first by right, rank, sense & habit in every Company; & prerogative becomes such a matter of course that nothing appears extraordinary that she does.[31]

The singularity and idiosyncrasy of Dashkova's behavior—as with the "degree of admiration" she assumes for herself—is defined by the "privileged sort of Mortal she makes herself"; that is, her claim to set the norm "by right, rank, sense & habit in every Company"—by right of her greater virtue. Martha also attributes Dashkova's sense of entitlement, of expected deference, to her "imperial habits" (i.e., she again underscores the connection Dashkova assumes between political privilege and her superior moral virtue). Elsewhere she also comments on the "singularity" of Dashkova's behavior in society:

> the compound of contradictions which form Princess D's character exceed belief. There are times when she is perfectly a Woman of fashion & very elegant in her manners, but she has learnt so little of the art of concealing her feelings, whatever they may be, that she often is settling according to her own fancy the dishes on the table at the moment that the guests are all waiting to eat them & a hundred other singularitys which it would be foolish & even wrong to write where they are so thoroughly counteracted by the admirable qualitys of her heart and understanding, by her invariable & comical love of truth (which makes her tell out things that set a large Company staring, twittering, blushing, biting their lips, and betraying a thousand different emotions *not one of which she ever remarks*), by her Celebrity, her rank & age, all which give her a right to be an Oddity, & Nature has stampt her such in the very fullest sense of the word.[32]

Here the "right to be an Oddity" suggests Dashkova's purposeful cultivation of celebrity, her playing on the notoriety and special privileged status such behavior implicitly bestowed. In any case, the same basic mechanism is at work, as her idiosyncratic behavior is as balanced or justified by "the admirable qualtys of her heart and understanding, by her invariable & comical love of truth." Paradoxically, the most marked *singularity* is founded on the conviction of supra-personal, *universally applicable* virtue. Whether speaking the truth to sovereigns at court or guests at table she frames her behavior in a way that highlights this special status. At the same time, she asserts both her independence from, and paradoxical reliance on, public approbation. Her silence in the face of the public reaction ("staring, twittering, blushing, biting their lips, and betraying a thousand different emotions"), which in confronting tyrants signals her moral untouchability, here suggests a game in which she can demonstrate her peculiar claim on social superiority. If in one context such self-positioning could make Dashkova "a Public Scourge" who bravely exposes evil, as Franklin had noted, in other contexts "such extream nicety," when perceived as "a kind of foppery in morals," might also easily invite ridicule.[33]

The Tragic Side: Allies and Traitors

Like a heroic "great soul" of classical tragedy, the protagonist is surrounded by a world that cannot possibly equal or appreciate her. In the confrontation with Emperor Paul, Dashkova predicts her imminent maltreatment: "one of the many prophecies I have made which have come true." She thus expresses both her superior understanding of the world, and, perhaps also on some deeper level, a comprehension that she sets the bar of virtue so high as to virtually invite persecution. Dashkova at one point describes herself as "an unhappy princess over whom a wicked wizard had cast an age-long spell,"[34] and this is an apt characterization of the "tragic" self-image that haunts her from childhood. The absolute terms in which Dashkova frames her life tends to turn the world into a huge conspiracy to frustrate her virtuous strivings. Dashkova casts herself in the role of victim—so pure, innocent, and virtuous that the world cannot help but be deficient and ungrateful. Failure, then, is not only inevitable but serves to confirm virtue and indeed reinforces the conviction of moral superiority. Hers is a Cassandra-like tragic self-consciousness, trapped in frustrated virtue that is both self-defeating and self-justifying.

As in the episode with Paul, Dashkova sees her life as an exalted moral struggle between good and evil, life and death. In this struggle, only a very chosen few are able to live up to her altruistic standards. One such ally is her husband, Mikhail I. Dashkov. His early death in 1764, leaving Dashkova with two young children, while one of those "bitter sorrows" that punctuated her existence also perhaps helped to solidify his ideal image in Dashkova's psychic economy.[35] Dashkova was twenty-one at the time of his death and never remarried. She describes their love as unconditional and all-encompassing, and tells a remarkable story of how she made a clandestine night visit to him at her mother-in-law's while pregnant (he was sick and trying to conceal this from both wife and mother). This episode strangely prefigures her conspiratorial behavior and dedication to Catherine.

Dashkova's relationship with the empress was paradoxical. On the one hand, Catherine embodied Dashkova's ideal of a public self, the person for whom she had risked her life, and on whom she staked her reputation and the good of the country. On the other hand, and perhaps due to this very idealization, their personal interactions after the "revolution" served as a continuing source of petty conflict and misunderstanding, the precise reasons for which are still often obscure. Catherine obviously provided Dashkova with her entrée into public life, both in accepting her services as a conspirator and later in appointing the princess to high public office. The empress also played a key formative role in Dashkova's personal development.[36] She was an inspiring role model—intellectually brilliant, charismatic, politically astute and ambitious—and her meteoric career, in which Dashkova took pride in having played a significant part, offered a powerful vindication of Dashkova's own ambitions in the public, overwhelmingly male arena. However, and starting immediately after Catherine's elevation to the throne, Dashkova found that the empress's behavior left many things to be desired. One constant area of friction was Dashkova's demand for Catherine's greater recognition of her selfless dedication and merit. Another perhaps related issue was Dashkova's disapproval of the empress's taking lovers ("favorites"), which was magnified by Daskova's early opposition to the political influence of the Orlovs and her sympathy for the party of Nikita I. Panin.[37]

Disguise, Concealment, Blindness

With a few exceptions, then, almost none of the people close to Dashkova could fulfill her exalted expectations. Most obviously, and most painfully for Dashkova, were those closest to her, her children. In their adult lives both son and her daughter miserably failed to live up to their mother's expectations. These disappointments, never fully explained, cast the most ominous shadow over her virtuous self-image. Like the twittering at table, but far more threatening, Dashkova acknowledges, and then purposefully ignores, those episodes that reveal the fragility, not to say immanent collapse, of her façade of unqualified virtue:

> Criticism and malicious gossip, which I could treat with contempt
> in the perfect confidence that I was acting as a good mother should,
> were not, unfortunately, the only sorrow that [the] marriage [of my
> daughter] brought me.

> But I am determined to pass over in silence the most bitter of all the
> unhappy experiences I have had in my life, and shall continue with
> my narrative.

> . . . if the sorrows which oppressed my heart were such that I should
> willingly have concealed them from myself, I could not now reveal
> them to the general public.[38]

In discussing her children, some have given credit to Dashkova for allegedly offering a validation of the female, private sphere, but it seems to me that Dashkova's image of motherhood belongs primarily to her virtuous, male, public self.[39] As we have seen, Dashkova denied the autonomous value of "private interest," and she also sees "motherhood" in terms of her disinterested service (to her children and to the public).[40] Dashkova here too asserts her unalloyed virtue in the face of the public "criticism and malicious gossip" that conspires with her children's disloyalty to challenge "the perfect confidence that I was acting as a good mother should." (In later life, the Wilmots—who encouraged her to write the memoir—took on the role of surrogate "family," offering Dashkova the security of unconditional veneration.) Characteristically, Dashkova preserves the image of her transparent virtue by an act of intellectual will, by refusing to confront certain issues, concealing them not only from "the general public" but perhaps also from herself. Ultimately, the ideal, or illusion, of total transparency can only be maintained by an act of purposeful blindness.

Dashkova's monolithic image of the virtuous self thus constantly threatens to unravel, and, as critics have alleged, there is a basic tension in the memoir between disguise and revelation, a discontinuity among the various "selves" that Dashkova never quite reconciles.[41] From this perspective, her self-presentation becomes a game of masks, a series of artificial, theatrical poses that do not necessarily cohere into a unified whole. One striking example that also exemplifies her play with gender roles is Dashkova's posture as "a simple old rustic"[42] woman whose naïve candor is sharply contrasted to the selfish intrigues of cosmopolitan court life. As opposed to the "tragic" persona described above, in which the heroine follows a strict code of "male" virtue (that of the "Great Man"), Dashkova at times also characterizes herself as a "Ninette at court." She is referring to the character made famous by a comic opera with that title by Charles Simon Favart,[43] in which a young countrywoman rejects the corrupting influence and amorous deviousness of court for the simple, honest pastoral life of the village. Of course, Dashkova herself—a complex, sophisticated, literate urbanite, and a princess schooled in high court intrigue—hardly fit the role of a simple peasant woman.[44] Yet if in the tragic role Dashkova cast herself as direct and confrontational, this one allowed for self-effacement and defensive retreat; notably, however, both roles feature an idealized virtuous heroine. As Martha Wilmot noted in the passages cited earlier, Dashkova's inability to disguise her feelings could suggest both a serious tragic side—her threatening potential as "a Public Scourge"—and also a "comical love of truth" based on her claiming "the right to be an Oddity." Dashkova's tragic and comic masks, then, are in some sense equivalent, although they nevertheless remain *masks*, incomplete reflections or disguises.

One moment when Dashkova assumes the mask of "female" country innocence is when Catherine offers her the position as first woman head of the Russian Academy of Sciences and she tries to refuse. Catherine's appointment triggers a minor crisis for Dashkova, who (somewhat uncharacteristically) fears herself unworthy. The two women engage in a peculiar negotiation of Dashkova's public stature, which hinges not only on Dashkova's qualifications but also on how the nomination will reflect on the empress's reputation. Catherine concludes on a paradoxical note, typical of a male hero narrative: "your refusal . . . has only confirmed my opinion that I could not have made a better choice."[45] Among the arguments Dashkova puts forward against her nomination is that "God himself, by creating me a woman, had exempted me from accepting the employment of a Director of an Academy of Sciences,"[46] perhaps the only time in the memoir that Dashkova disparages capability purely on grounds of gender. Dashkova expresses amazement at "the extraordinary step you have just taken in making me *Monsieur le Directeur* [author's italics] of an Academy of Sciences" and warns the empress "that you will soon tire of leading the blind, for indeed I shall be an ignoramus at the head of Science."[47] She underscores her ignorance and inability by describing herself as blind. However, all this is but a prelude to the resounding success of Dashkova's powerful, virtuous, intellectual "male" persona, as Dashkova's directorship takes the Academy to a new level of prosperity and achievement.[48] As one critic has noted about her initial refusal, Dashkova "makes protestations of incapacity and modesty, even as she details her capabilities and accomplishments."[49] Dashkova overcomes her reticence (or the specter of false modesty?) in assuming the directorship by arranging to be presented to the assembled academicians in her new office by the great mathematician Leonhard Euler. Significantly, perhaps, is the fact that Euler had already become *blind* by this time. Dashkova thus achieves visibility and prominence despite—or perhaps by virtue of— her self-proclaimed weakness and "blindness," which paradoxically turn out to signal her own status, comparable to that of Euler, as a "Great Man."

Indeed, Dashkova repeatedly emphasizes the complete transparency of her motives. In a key passage in which she defends her relationship with Catherine, she asserts the total openness, the visible virtue, of her writing:

> I want to disguise nothing in this narrative. I shall tell of the little
> differences that cropped up between Her Majesty and myself, hiding
> nothing; and the reader will see for himself that I never fell into
> disgrace, as has been claimed by several writers who wanted to harm
> her interests, and that if the Empress did not do more for me, it was
> because she had an intimate knowledge of me and was quite aware
> that every form of self-seeking was entirely alien to my nature; besides,
> and without false modesty, in the midst of Court life, I revealed a
> heart so artless, so unspoilt, that I forgave even those who showed
> black ingratitude, egged on as they were by my all-powerful enemies
> who managed to turn against me those to whom I had rendered
> great services. Nevertheless, forty-two years have passed before I have
> ventured to reveal the whole of my experience of human ingratitude,

which, however, never made me tired of doing all the good of which I was capable, often at the cost of great financial inconvenience, for my means were more than modest.[50]

This remarkable passage offers a useful summation of the workings of Dashkova's self-image. On the one hand, she offers herself as totally virtuous, disguising nothing, someone for whom "every form of self-seeking was entirely alien"; her heart is pure, "artless," "unspoilt," and stoically forgiving. She will tell of the "little differences" she had with the empress (because there were no big ones), and "hiding nothing . . . the reader will see for himself." Yet this mask of stoic virtue and all-forgiveness is immediately undercut by the fact that the entire reason for writing the memoirs is precisely *to get back her own*, to set the record straight, to reveal "the whole of my experience of human ingratitude," stored up over the course of forty-two years; that is, from the time of Catherine's ascension to the throne in 1762 to the time Dashkova finished writing in 1805. This latter, vehemently self-righteous and hyperbolically defensive posture undercuts the pose of artless simplicity, forcing the reader (especially perhaps a modern critical one) to take her pronouncements with a grain of skepticism. From Dashkova's perspective, though, the writing of the text is motivated by the conviction that "virtue must advertise"—if only after holding back for forty-two years. Writing a memoir offered a magnificent opportunity to have the last word.

Notes

1. Historians still debate Dashkova's assertion of her key role. Catherine herself—to Dashkova's dismay—disparaged her participation immediately following the coup, which may have been for political reasons, insofar as Dashkova hoped to counter the Orlovs' influence.

2. E. R. Dashkova, *The Memoirs of Princess Dashkova*, trans. and ed. Kyril Fitzlyon (Durham, 1995), p. 248. Unfortunately, there still is no fully authoritative version of Dashkova's memoirs, which exist in two basic variants. On the history of the problem, see A. Woronzoff-Dashkoff's "Afterword" in Dashkova, *Memoirs*, pp. 284–89, and his "Additions and Notes in Princess Dashkova's *Mon histoire*," *Study Group on Eighteenth-Century Russia Newsletter*, vol. 19 (1991), pp. 15–21. See also the recent composite French text, which lacks a critical apparatus: Princesse Dachkova, *Mon histoire: mémoires d'une femme de lettres russe à l'époque des Lumières*, ed. Alexander Woronzoff-Dashkoff, Catherine Le Gouis, and Catherine Woronzoff-Dashkoff (Paris, 1999).

3. Dashkova, *Memoirs*, p. 279.

4. A. Woronzoff-Dashkoff, "Disguise and Gender in Princess Dashkova's *Memoirs*," *Canadian Slavonic Papers*, vol. 33 (1991), p. 62.

5. Dashkova, *Memoirs*, pp. 271–72. On the problem of the empress's reputation, and Dashkova's role in "the cult of Catherine," see Simon Dixon, "The Posthumous Reputation of Catherine II in Russia, 1797–1837," *Slavonic and East European Review* 77:4 (1999): 646–79.

6. On European debates over Russia's status, see Dimitri S. von Mohrenschildt, *Russia in the Intellectual Life of Eighteenth-Century France* (1936; reprint, New York, 1972); Albert Lortholary, *Le Mirage Russe en France au XVIIIᵉ siecle* (Paris, [1951]); Isabel de Madariaga, "Catherine and the *Philosophes*," in *Russia and the West in the Eighteenth Century*, ed. Anthony Cross (Newtonville, Mass., 1983), pp. 30–52; and Larry Wolff, *Inventing Eastern Europe: The Map of Civilization on the Mind of the Enlightenment* (Stanford, 1994). On Russia's

problematic status in earlier writings, see esp. Marshall Poe, *"A people born to slavery": Russia in Early Modern European Ethnography, 1476–1748* (Ithaca, 2000).

7. The quotation is taken from the memoirs' dedicatory letter to Martha Wilmot, not included in the Fitzlyon translation or 1999 French edition. I cite it from: E. R. Dashkova, *Zapiski. Pis'ma sester M. i K. Vil'mot iz Rossii,* ed. S. S. Dmitriev, comp. G. A. Veselaia (Moscow, 1987), p. 35. See also Dashkova's mention of these libels in Dashkova, *Memoirs,* pp. 51, 62, 91, 279.

On Dashkova's autobiography in the broader literary context of "Russica," see Kelly Herold, "Russian Autobiographical Literature in French: Recovering a Memoiristic Tradition (1770–1830)," (Ph.D. diss., University of California, Los Angeles, 1998). On the particular writers Dashkova repudiates, see V. A. Somov, "'Prezident trekh akademii': E. R. Dashkova vo frantsuzskoi 'Rossike' XVIII veka," in *E. R. Dashkova i A. S Pushkin v istorii rossii,* ed. L. V. Tychinina (Moscow, 2000), pp. 39–53. On Russians' familiarity with "Russica," see Somov's "Frantsuzskaia 'Rossika' epokhi prosveshcheniia i russkii chitatel'," in *Frantsuzskaia kniga v Rossii v XVIII v.: Ocherki istorii,* ed. S. P. Luppov (Leningrad, 1986), pp. 173–245.

8. Dashkova, *Zapiski,* p. 36.

9. Dashkova, *Memoirs,* pp. 33, 34.

10. Martha and Catherine Wilmot, *The Russian Journals of Martha and Catherine Wilmot, 1803–1808,* eds. Marchioness of Londonderry and H. M. Hyde (London, 1934), p. 211. Farmer General of the Empire was something like a minister of finances or the official in charge of taxes and revenue.

11. Judith Vowles contrasts Catherine the Great's ability to reconcile "the claims of worldly society and the intellectual life" to Dashkova's rejection of "feminine" social pursuits (e.g., the life of the salon) in favor of serious "male" interests. See "The 'Feminization' of Russian Literature: Women, Language and Literature in Eighteenth-Century Russia," in *Women Writers in Russian Literature,* eds. Toby W. Clyman and Diana Greene (Westport, Conn., 1994), pp. 40–44.

12. Dashkova, *Memoirs,* p. 35.

13. Arthur O. Lovejoy, *Reflections on Human Nature* (Baltimore, 1961). See also Monika Greenleaf's discussion of Catherine the Great's memoirs, and especially the idea—connected with Adam Smith—of "the creation of [an autobiographical] private life as a spectacle designed to elicit sympathy in others." "Performing Autobiography: The Multiple Memoirs of Catherine the Great (1756–96)," *Russian Review,* vol. 63 (2004), p. 412. Dashkova, of course, had close personal and intellectual connections to Smith and the Scottish Enlightenment, but these have yet to be investigated.

14. I discuss various aspects of the primacy of the visual in eighteenth-century culture in: "The 'Obviousness' of the Truth in Eighteenth-Century Russian Thought," in *Filosofskii vek,* 24: *Istoriia filosofii kak filosofii,* Chast' 1 (St. Petersburg, 2003), pp. 236–45; "Dialektika videniia v *Puteshestvii* Radishcheva," in *A. N. Radishchev i evropeiskoe Prosveshchenie: Materialy Mezhdunarodnogo simpoziuma, 24 iiunia 2002 g.* (St. Petersburg, 2003), pp. 36–47; "Oda kak otkrovenie: O pravoslavnom bogoslovskom kontekste lomonosovskikh od," in *Slavianskii almanankh 2003* (Moscow, 2004), pp. 368–84; and "'Vechernee' i 'Utrenee razmyshleniia o Bozhiem velichestve' Lomonosova kak fiziko-teologicheskie proizvedeniia," in *XVIII vek,* vol. 24 (St. Petersburg, forthcoming).

15. Her famous *Instruction* (*Nakaz*) to the delegates assembled by her order to draft a new law code for Russia offers a dramatic expression of this view. See *Documents of Catherine the Great: The Correspondence with Voltaire and the Instruction of 1767 in the English text of 1768,* ed. W. F. Reddaway (1931; reprint, New York, 1971), and other editions. On Catherine's quest for visibility, see David M. Griffiths, "To Live Forever: Catherine II, Voltaire, and the Pursuit of Immortality," in Roger Bartlett et al., *Russia and the World of the Eighteenth Century* (Columbus, Ohio, 1988), pp. 446–68; Simon Dixon, *Catherine the Great* (New York, 2001), chapter 3; and Greenleaf, "Performing Autobiography."

16. Dashkova, *Memoirs,* p. 251. Here and below I have changed "Pavel I" to "Paul I."

17. Dashkova herself makes the comparison that "my life could serve a subject for a heartrending novel" (Dashkova, *Zapiski,* p. 35), and the memoir is punctuated with theatrical terms (tragedy, farce, comedy, the stage, etc.). These sorts of references may perhaps be common in most autobiographical writing, but my suggestion is that Dashkova shared the special self-image and discourse about virtue and self-display which

were reflected in Russian classicist literary works, whose very function was to offer Russian society a "school for virtue."

18. Dashkova, *Memoirs,* pp. 262 and 264.

19. Wilmot, *Russian Journals,* pp. 55–56.

20. Catherine Wilmot likewise commented in a letter to Anna Chetwood that "Three thousand Peasants, 'my subjects' (as she calls them) live most happily under her absolute power; and of all the blessed hearted beings that ever existed on that subject she is the most blessed (excepting your Mother)." Wilmot, *Russian Journals,* p. 199.

21. M. M. Safonov, "Ekaterina malaia i ee 'Zapiski,' " in *Ekaterina Romanovna Dashkova: issledovaniia i materialy,* in A. I. Vorontsov-Dashkov et al. (St. Petersburg, 1996), p. 21. However, Dashkova did not or would not admit to any contradiction between Catherine's regime and the moral imperative, although as Safonov's argument suggests, it was not too far a distance from Dashkova's "courage to be a personality" to the appearance of Decembrist revolutionary ferment.

22. Compare Richard Wortman's description of Gavriil R. Derzhavin's memoirs: "Their most striking characteristic for the historian … is Derzhavin's ego, his limitless confidence in himself, the wonderful naïve sense that his personal progress and success are identical to the cause of justice and the national well-being. This boundless self-certainty, which would be lacking in memoirs of a later era, provides the central unity and verve of the *Zapiski*" (Richard Wortman, "Introduction," *Perepiska (1794–1816) i "Zapiski"* [1871; reprint, Cambridge, 1973], pp. 2–3).

Writers' insistence on equating personal and universal merit was also a central problem in establishing the norms of literary usage, which made literary critical discourse of mid-century Russia notoriously acrimonious. See my discussion in "Paskvil', polemika, kritika: 'Pis'mo … pisannnoe ot priiatelia k priiateliu' (1750g.) Trediakovskogo i problema sozdaniia russkoi literaturnoi kritiki," *XVIII vek,* vol. 21 (St. Petersburg, 1999), pp. 62–72.

23. Benjamin Franklin, *The Autobiography of Benjamin Franklin,* intro. Lewis Leary (New York, 2004), p. 74.

24. Franklin, *Autobiography,* p. 66.

25. Franklin, *Autobiography,* p. 75.

26. Franklin, *Autobiography,* p. 2.

27. Franklin, *Autobiography,* p. 73.

28. Franklin describes the efforts he made (and the "some violence" required) to control his "natural inclination" to express his opinions in confident and categorical terms, a moderation that with time he says became habitual. (See for example Franklin, *Autobiography,* p. 64). On Franklin's self-control and phenomenally successful pursuit of approbation, see Edmund S. Morgan, *Benjamin Franklin* (New Haven, 2003), especially chapter 1.

29. "O istinnom blagopoluchii," *Sobsednik liubitelei rossiiskogo slova,* vol. 3 (1783), pp. 24–34; my citation is from E. R. Dashkova, *O smysle slova "vospitanie": sochineniia, pis'ma, dokumenty,* ed. G. I. Smagina (St. Petersburg, 2001), p. 130.

30. Dashkova, *Memoirs,* p. 276.

31. Wilmot, *Russian Journals,* p. 196.

32. Wilmot, *Russian Journals,* p. 360. Italics in the original.

33. Indeed Dashkova is highly sensitive to becoming (as she puts it) the "dupe of my own conscientious scruples" (Dashkova, *Memoirs,* p. 198), which she feels happening quite often.

34. Dashkova, *Memoirs,* p. 242.

35. And as Kyril Fitzlyon notes, they spent much of their short married life apart (Dashkova, *Memoirs,* p. 305).

36. Dashkova describes "earning the esteem" of the then Grand Duchess Catherine as a turning point in her young life. Catherine, who shared her intellectual passion (Dashkova notes that she was the only other woman of her day "who did any serious reading"), captured her "heart and mind," satisfying the emptiness that her privileged home education had failed to fill (Dashkova, *Memoirs,* pp. 32, 35–36).

37. Dashkova's dislike of favoritism reflected her moral condemnation of "private" values that

contradicted the public good (discussed below). This ethical aversion probably also had a more personal component, as something that also challenged her own role of widowed celibacy. Her more nasty critics alleged her ambiguous sexuality (that of an Amazon or hermaphrodite; see Derzhavin's couplet on Dashkova: *"K Portretu Germafrodita: 'Se lik: / I baba i muzhik'"* ["To the Portrait of a Hermaphrodite: 'This face / Is both a *baba* and *muzhik*'"], in G. R. Derzhavin, *Sochineniia,* ed. Ia. Grot, [St. Petersburg, 1870], vol. 3, pp. 270–71. My thanks to V. Proskurina for this reference).

As far as Dashkova's possible political opposition to Catherine, she, like Nikita I. Panin, was an advocate of "limited monarchy" (Dashkova, *Memoirs,* p. 60); that is, a limitation of Catherine's autocracy via aristocratic power sharing, although she did not seem to approve the Swedish model that Panin promoted (Dashkova, *Memoirs,* pp. 65, 67). Dashkova's political conflict with Catherine, however, remains obscure and should not be overstated. On Panin's program, see David L. Ransel, *The Politics of Catherinian Russia: The Panin Party* (New Haven, 1975), esp. pp. 112–13. As noted earlier, M. M. Safonov sees Dashkova's conflict with Catherine primarily in political terms, rising out of the nature of autocracy (see n. 21).

38. Dashkova, *Memoirs,* pp. 143 and 280.

39. Barbara Heldt argues the former position, and sees a "blending" of Dashkova's public and private selves, arguing that a "balance, the classical symmetry she seeks, is almost never realized at any one time," but emerges "over a lifetime." See her *Terrible Perfection: Women and Russian Literature* (Bloomington, 1987), pp. 69–71.

40. Characteristically, Dashkova has various prominent public figures giving voice to this view, as when the queen of England declares that "I have always known . . . that there are few mothers like you." (Dashkova, *Memoirs,* p. 151).

41. This is close to the position of A. Woronzoff-Dashkoff ("Disguise and Gender," p. 63, partially repeated in the "Afterword" to *Mon histoire*), who foregrounds the concealment and dissimulation in Dashkova's memoirs and argues: "Dashkova's tragedy was that she could not realize her dreams and desires within the accepted norms of eighteenth-century female behavior." Woronzoff-Dashkoff puts special emphasis on the episode in the autobiography when Dashkova and companions repaint a series of paintings in a Danzig hotel in order to turn a Prussian victory into a Russian one—an episode that suggests Dashkova's "dissimulation" to the point of positive falsification, however much inspired by a carnivalesque spirit of masquerade. This approach seems to put the emphasis in the wrong place, as it undervalues Dashkova's fundamental desire for transparency and validation on her own merits.

42. Dashkova, *Memoirs,* p. 156.

43. Favart's *Ninette à la Cour, ou Le caprice amoureux* (1755) was a parody of the very popular opera *Bertoldo in corte* or *Bertoldo alla corte* (originally titled *Bertoldo, Bertoldino e Cacasenno*), libretto by Carlo Goldoni and music by Vincenzo Ciampi (1748). See *The New Grove Dictionary of Music and Musicians,* 2nd ed. (London, 2001), vol. 5, p. 830 and vol. 8, p. 623; and O. G. Sonneck, "Ciampi's 'Bertoldo, Bertoldino e Cacasenno' and Favart's 'Ninette à la Cour,'" in his *Miscellaneous Studies in the History of Music* (New York, 1921), pp. 111–179.

My attention was drawn to this issue by Lyubov Golburt, "Discourses of the Self in the Eighteenth-Century Russia: E. R. Dashkova's *Mon Histoire*," delivered at the AATSEEL National Convention, New York, Dec. 28, 2002 (for the abstract see http://www.aatseel.org/program/aatseel/2002/abstracts/Golburt.html, accessed February 23, 2005). Goubert argues that Dashkova's goal was "to portray herself not as just another court lady, but as a distinct public figure. In addition, pretending to aspire to rustic bliss, Dashkova . . . flaunts the conventional, pastoral values for the sake of her autobiographical reliability."

44. O. G. Sonneck remarks that for Favart's character the innocent pose was itself already more or less of a pretense. He notes that Goldoni's "naïve and somewhat primitive but shrewd and quick-witted Italian peasant woman . . . has become in Favart's hands a typically Parisian *villgeoise,* equally quick-witted but no longer naïve or primitive. Far from it, she has turned into a very dexterous type of stage coquette." (Sonneck, "Ciampi's 'Bertoldo,'" p. 165).

45. Dashkova, *Memoirs,* p. 210.

46. Dashkova, *Memoirs,* p. 201.

47. Dashkova, *Memoirs*, p. 204.

48. See Michael Gordin's essay in this volume.

49. Vowles, "The 'Feminization,'" p. 44.

50. Dashkova, *Memoirs*, p. 96; I have edited the passage to bring it closer to the French (*Mon histoire*, p. 68).

A Man Made to Measure:

Benjamin Franklin, American Philosophe

KAREN DUVAL

*T*he unlikely meeting of Benjamin Franklin and Princess Ekaterina Romanovna Dashkova (1743–1810) in Paris, early in 1781, brought together two unusual figures from opposite sides of the world, separated by age, experience, and rank. By all rights of tradition the princess's presence in that city was to be expected and celebrated; but, Franklin's, a printer from Philadelphia, was not. And yet Dashkova, reputed to have participated in the coup that brought Catherine II to the throne, and possessed of an almost mannish appearance, was regarded more often as an object of curiosity than as someone to know. By contrast, Franklin was idolized in Paris, and his simple manner and modest costume, fashioned along the lines of his character, were very much suited to his purpose in the French capital: to be the spokesman for his country and to embody its ideals. That he came to be there in that role was as unlikely as his meeting with the princess.

Born far from the capitals of Europe, on the other side of the Atlantic, the youngest son of a tallow chandler and soap boiler, Franklin had little formal education. What schooling he did have was completed by age ten. He spent his early years as an apprentice in his brother James's print shop in Boston, where an atmosphere of freethinking and access to a wide variety of books were the school that his father had not been able to afford him. A chance copy of Addison and Steele's *Spectator*[1] became his composition primer. Delighted by the tone and form of the essays, he studied them closely and tried his hand at imitating them. He would turn some of the prose tales into verse and then, after a lapse of time and memory, turn them back again into prose, striving for the variety and arrangement of the originals. In this way he gradually developed a style that was "familiar but not coarse, and elegant but not ostentatious."[2] His fourteen "Silence Dogood" essays, printed anonymously in James's *New-England Courant,* are the fruit of the lessons Franklin learned in his brother's print shop. Inspired by the *Courant's* satirical essays and modeled generally on those of the *Spectator*, these writings are "letters to the editor" written in the guise of a minister's widow, who comments on a range of topics, including the natural equality of women.[3]

Franklin pursued his self-education in the books he found to hand and in those he was able to borrow or buy. From his early reading of his father's "Books in polemic Divinity," Franklin acquired a taste for disputation, an art he later perfected by studying Antoine Arnauld and Pierre Nicole's *Logic, or the Art of Thinking*.[4] In the work of Greek historian Xenophon[5] he discovered the Socratic method and took an adolescent's delight in using it to confound his friends as they debated matters large and small. With time Franklin adopted more subtle forms of persuasion, but his study of Socrates had instilled in him a habit of picking up a question and turning it around in the light of reason. Among the many books he read during this period was John Locke's *Essay Concerning Human Understanding,* from which he learned that truth could be grasped from direct experience. The Calvinist beliefs that he had grown up with were unsettled by his reading of Lord Shaftesbury, a religious skeptic, and by Anthony Collins, a deist.[6] Soon Franklin's own "indiscrete Disputations about Religion" caused a stir amongst the townspeople who took him to be an infidel or an atheist. This unwanted attention, together with his brother's "tyrannical Treatment" of him led the seventeen-year-old "to assert [his] Freedom" by running away to Philadelphia.[7] Although at this point he was still two years away from completing his apprenticeship as a printer, he had already acquired the skills of his trade, and through his passionate reading and dedication to self-improvement Franklin had achieved an intellectual mastery of some of the basic "text books" of the Enlightenment.

Franklin's flight from Boston opened to him a world of experience that would shape the lessons learned in his early studies. But for the next four years, in Philadelphia and then in London, he lived without direction and without a plan. Indeed, as described in *The Autobiography,* those years had a picaresque and serendipitous quality about them. A chance encounter in Philadelphia brought Franklin to the attention of Sir William Keith, governor of Pennsylvania, who saw in him a "young Man of promising Parts."[8] Proposing to set Franklin up as a printer, Keith sent him off to London with promises of letters of introduction and credit. When these failed to arrive, Franklin found work at Samuel Palmer's printing house, where he was given the job of typesetting an edition of *The Religion of Nature Delineated,* William Wollaston's treatise against the rational morality of the deists. The compositor in Franklin soon gave way to the disputer; he wrote and printed *A Dissertation on Liberty and Necessity, Pleasure and Pain*, his rebuttal to Wollaston's book based on his readings of Locke, Shaftesbury, and Collins. Views that had caused a scandal in Boston opened doors to him in London. William Lyons, a surgeon and author of *The Infallibility, Dignity, and Excellency of Human Judgment,* read Franklin's pamphlet, sought him out, and introduced him to a number of coffeehouse clubs and his circle of intellectual friends, among them Bernard Mandeville, the Dutch physician and author of *The Fable of the Bees,* and Henry Pemberton, another physician and writer who had ties to Isaac Newton. As always, Franklin read eagerly and constantly, he attended plays, and he talked. He might have gone on living in London like this indefinitely, leaving the promise of his own talents and skills to chance and circumstance, if a Quaker merchant named Thomas Denham had not persuaded him to return to Philadelphia in his employ. So after eighteen months abroad, with no increase in his fortune other than having "pick'd up some very ingenious Acquaintance whose Conversation was of great Advantage" and "read considerably," Franklin returned to Philadelphia, not yet twenty-one years old.[9]

Franklin's passage back across the Atlantic marked the end of his apprenticeship, and his return to Philadelphia marked a true beginning. Life in London had been more stimulating for Franklin than anything he had known before, but it was back in America, on the grid-like streets of Philadelphia, that he would find the freedom and the means to exercise his talents more fully. To do so, however, he would have to proceed differently: "I have never fixed a regular design in life; by which means it has been a confused variety of different scenes. I am now entering upon a new one: let me, therefore, make some resolutions, and form some scheme of action, that, henceforth, I may live in all respects like a rational creature."[10]

The man Princess Dashkova met fifty-five years later had fully realized the plans he set for himself on that ocean crossing in 1726. He had shaped a life according to the belief "that the most acceptable Service of God was the doing Good to Man."[11] He had made a life for himself as a printer in Philadelphia, using his press to inform and enlighten his readers.[12] In countless ways he had contributed to the civic development and political life of his city and country. Having been denied the structure of a formal education in his own youth, he had created an array of educational institutions in his adoptive city, many of them undertaken to further his own desire to learn. The Junto was the first of these, his "Club for mutual Improvement," where questions of morals, politics, and science were discussed and examined as in a coffeehouse with more structure and order: "Our Debates were to be under the Direction of a President, and to be conducted in the sincere Spirit of Enquiry after Truth, without Fondness for Dispute, or Desire of Victory." [13] The Library Company, which grew out of the members' need to supplement their separate personal libraries, was Franklin's first public project. [14]

Franklin's greater involvement in projects for the public good gradually led him into projects of a political nature, and early in 1757 he accepted the position to serve as the agent of the Pennsylvania Assembly to England. Delay followed delay, but Franklin and his son, William, eventually sailed for England on June 23 of that year. [15] The crossing itself was swift, and gave Franklin occasion to wonder about a ship's speed relative to the shape of the hull, or the dimensions and placement of the mast, or the shape of the sails. [16] He and his son arrived in London on July 27, having stopped on their way to view Stonehenge.

Franklin's early efforts in London as a colonial agent tended toward an amicable and equitable resolution of disputes between the colonies and Great Britain. But in 1768 he observed that those efforts were often undermined by his ties to both countries: "Being born and bred in one of the countries, and having lived long, and made many agreeable connections of friendship in the other . . . I do not find that I have gained any point in either country, except that of rendering myself suspected by my impartiality; in England of being too much an American, and in America of being too much an Englishman." [17]

No such conflict existed for Franklin in France. On the contrary, Franklin represented his country to a much more sympathetic audience. Indeed, Franklin's reputation as a scientist was established in France before he received any great public recognition in England for his scientific work. Although his papers on electricity were reviewed by some members of the Royal Society of London in 1747, only extracts were printed the following year in the Society's *Philosophical Transactions*. Publication in full did not occur until three years later when Peter Collinson had Franklin's work printed in London "under the Inspection and Correction of [their] Learned and Ingenious Friend Dr. Fothergill." The work, which contained the first suggestion that the electrical nature of lightning could be verified experimentally, finally appeared in April 1751, in pamphlet form, with the title *Experiments and Observations on Electricity, made at Philadelphia in America*. [18] The publication quickly caught the attention of the French naturalist Georges-Louis Leclerc, comte de Buffon, who urged Thomas-François Dalibard, a botanist and physicist, to translate the work. *Expériences et Observations sur l'électricité faites à Philadelphie en Amérique par M. Benjamin Franklin; & communiquées dans plusieurs Lettres à M. P. Collinson de la Société Royale de Londres* appeared in February 1752. Three months later Buffon and Dalibard, together with electrical experimenter Delor, performed the experiment that Franklin had sketched out to "determine the Question,

Whether the Clouds that contain Lightning are electrified or not." Franklin had suggested that a sentry box, big enough to accommodate a man and an electrical stand, be set on a high place. From the middle of the stand an iron rod was to be attached and, bending out of the door of the box, to rise twenty or thirty feet into the air so that it would attract sparks of electricity from a storm cloud passing overhead.[19] The experiment, performed with some modifications, was a success, and French enthusiasm for the American whose theories had just been proved was immediate and widespread. Within three days, Dalibard reported on the experiment to the French Royal Academy of Sciences; and King Louis XV, who had witnessed the demonstration, sent his compliments to Franklin care of the Royal Society in London.[20]

In 1766, not long before his first visit to France, Franklin was called before the British House of Commons to answer questions regarding colonial opposition to the Stamp Act. In the course of the examination Franklin presented a strong defense of American rights, which, in the opinion of some, was responsible for the repeal of the Stamp Act.[21] Franklin's testimony, published immediately in London and later in Paris in French translation, added to his growing fame and defined for many readers the essence of American political thought. Franklin became in the minds of Europeans generally, and the French especially, the defender of America's political rights and his country's principal spokesman.[22]

Franklin's visits to France in 1767 and 1769 gave substance to the myths that had grown up around the "electric Franklin." He established ties of friendship both personal and institutional, and saw to it that Frenchmen were elected to the American Philosophical Society, beginning with Buffon in 1768. In 1772 Louis XV himself nominated Franklin to a position of foreign associate in the French Academy of Sciences.[23] Dalibard continued to publish translations of Franklin's scientific work, and the two men corresponded regularly, exchanging news and books.

Recognizing that France, and publications in French, had for a long time played a central role in the spread of political and philosophical ideas, Franklin enlisted Jacques Barbeu-Dubourg, physician and *philosophe,* to translate John Dickinson's *Letters from a Farmer in Pennsylvania*[24]—a work that laid out the American position following the Stamp Act crisis and, in Franklin's estimation, played an important role in shaping public opinion in Europe:

> All Europe is attentive to the Dispute between Britain and the
> Colonies; and I own I have a Satisfaction in seeing that our Part is
> taken every where; because I am persuaded that that Circumstance
> will not be without its Effect here in our Favour. . . . In France they
> have translated and printed the principal Pieces that have been written
> on the American Side of the Question; and as French is the political
> Language of Europe, it has communicated an Acquaintance with our
> Affairs very extensively.[25]

In 1773 Barbeu-Dubourg published the *Oeuvres de M. Franklin,* the first collection of Franklin's writings in any language. This work, in two volumes, was especially welcome to the French reader, who was eager for anything by Franklin. In addition to Franklin's earlier scientific writings, Barbeu-Dubourg included pieces that had never appeared in print before, as well as extracts from his own correspondence with Franklin.[26] In a general preface Barbeu-Dubourg described Pennsylvania, as Voltaire had done in his *Lettres philosophiques* (1734), as a place where religious freedom was extended to people of all faiths, where virtue was preferred to theology, and where a spirit of equality allowed all people to prosper.[27] To accompany the portrait of Franklin on the frontispiece, Barbeu-Dubourg composed a quatrain that captured the range of this American *philosophe* as seen by the French at the time:

> Il a ravi le feu des cieux,
>
> Il fait fleurir les arts en des climats sauvages.
>
> L'Amerique le place à la tête des sages,
>
> La Grece l'auroit mis au nombre de ses Dieux.[28]

The publication of this translation of Franklin's work secured his place in the French pantheon of Enlightenment *philosophes.* Barbeu-Dubourg accurately predicted that Franklin's *Oeuvres* would also enhance the reputation of America in France and throughout Europe.[29]

Franklin recognized that America's image abroad sometimes bore little relation to reality. He had once commented wryly to a friend after reading "a Piece of Voltaire's on the Subject of Relegious Toleration . . . at a Time when we are torn to Pieces by Factions religious and civil, shows us that while we sit for our Picture to that able Painter, tis no small Advantage to us, that he views us at a favourable Distance."[30] He also understood how large a role that image could play and, in 1776, how crucial it could be to the success or failure of the country's newly declared independence. In that year it became Franklin's mission to represent his country in France and, as it were, to "sit for a portrait."

A measure of the degree to which Franklin had come to embody the idea of the American is best demonstrated by comparing his 1767 visit to France, when his "Taylor and Peruquier had transform'd [him] into a Frenchman . . . in a little Bag Wig and naked Ears," to his return in 1776, when a costume of unadorned simplicity was perfectly suited to the role that only he could play. In his simple garb, "an old Man with grey Hair appearing under a Martin Fur Cap, among the Powder'd Heads of Paris,"[31] Franklin personified the *philosophe* from America and was greeted with wild enthusiasm everywhere he went. He was Voltaire's "good Quaker" and Franklin's own "poor Richard" combined.[32] But, in the eyes of the French and much of Europe, Franklin had gone further and done something no other *philosophe* had dared to do, namely, to put theory into practice and declare independence.[33] In Anne-Robert-Jacques Turgot's epigrammatic rewriting of Barbeu-Dubourg's quatrain, Franklin was the man who seized lightning from the heavens and the scepter from tyrants (*Eripuit cœlo fulmen, sceptrumque tyrannis*).[34]

Franklin's return to France on behalf of his newly independent nation in 1776 also symbolized in some sense a return of the Enlightenment project to its source. As if by design, and after decades spent in exile, Voltaire also returned to Paris, just four days after the signing of the treaties of alliance between France and America. Franklin called on Voltaire almost immediately, and in April the two men appeared together publicly at the French Academy of Sciences. The sight of these two figures who in distant corners of the Enlightenment world had dedicated their lives to the increase of knowledge and the happiness of mankind, so moved the members that they called on the two men to embrace *à la française*. John Adams recorded the moment with his usual acerbity: "The two Aged Actors upon this great Theatre of Philosophy and frivolity then embraced each other by hugging one another in their Arms and kissing each others cheeks, and then the tumult subsided."[35] When Voltaire succumbed to exhaustion and illness at the end of May, Franklin played a prominent role in the memorial held in his honor by the Masonic Lodge of the Nine Sisters, into which both men had been inducted that spring.[36]

Franklin supplemented his official diplomatic activities with the kind of work he had always done when he embarked on a public project: he promoted it with every means at his disposal, letting his own accomplishments and prestige speak for those of his country. Together with the duc de La Rochefoucauld and others, Franklin ensured that news about America and the war circulated in the newspapers of France and Europe. He saw to it that the state constitutions were translated and published when they became available. Inhibited somewhat in formal settings by his lack of facility with spoken French, he nonetheless participated in meetings of the French Academy of Sciences, serving on several investigative commissions.[37] Although he no longer had time for scientific experimentation, he did occasionally "[throw] a few Thoughts on Paper," some of which were either read for him to the French Academy of Sciences or published in a journal. He remained interested in the work of others who had more time for experimental science and wrote to colleagues, assuring them if they had "lately published any new Experiments, or Observations, in Physicks, [he should] be happy to see them."[38] He exchanged letters with Jan Ingenhousz, answering the Dutch physician's questions about electrical phenomena and formulating experiments to test the conductivity of heat in metals—the procedure known now as the Ingenhousz experiment.[39] And, of course, he was present when the first balloon ascended in Paris in the summer of 1783. Ever alert to the possibilities of discovery, Franklin wrote to Sir Joseph Banks, president of the Royal Society, of this "Experiment of a vast Globe sent up into the Air," anticipating that "if prosecuted may furnish Means of new

Knowledge."[40] On August 27, a week before the final peace treaty was signed, Franklin witnessed the repetition of "the new aerostatic Experiment, invented by Messrs. Mongolfier of Annonay" by "Mr. Charles; Professor of Experimental Philosophy at Paris." He wrote again to Banks, describing at length the globe, the weather conditions, the crowds, and some of the remarks occasioned by the event.[41] When one skeptic questioned the good of such an invention, Franklin himself famously replied, "What good is a new-born baby?" (*Eh, à quoi bon l'enfant qui vient de naître?*).[42]

Franklin participated in the cultural life of the city, and with his friends in Passy he created a domestic life of easy and frequent socializing. He was ubiquitous in Paris, and so was his portrait. As he wrote to his daughter:

> The clay medallion of me . . . with the pictures, busts, and prints, (of which copies upon copies are spread every where) have made your father's face as well known as that of the moon, so that he durst not do any thing that would oblige him to run away, as his phiz would discover him wherever he should venture to show it. It is said by learned etymologists that the name *Doll,* for the images children play with, is derived from the word Idol; from the number of *dolls* now made of him, he may be truly said, *in that sense,* to be *I-doll-ized* in this country.[43]

When asked at a later date for his image, Franklin exclaimed: "I have at the request of friends sat so much and so often to painters and Statuaries, that I am perfectly sick of it. . . . I would nevertheless do it once more to oblige you if it was necessary, but there are already so many good Likenesses of the face, that if the best of them is copied it will probably be better than a new one. . . . The face Miss Georgiana has, is thought here to be the most perfect."[44]

On January 21, 1781, around the time of the Princess Dashkova's arrival in Paris, the *Journal de Paris* announced for sale an engraving of the "Minister plenipotentiary of the Republic of the United Provinces of North America," in which the cynic philosopher displays a portrait of Franklin.[45] The platform on which the portrait rests carries an inscription declaring that Diogenes has found the object of his search at last: *Stupete gentes! Reperit virum Diogenes* (Behold nations! Diogenes has found a man).[46] Diogenes, who even in the brightness of day had searched unsuccessfully for an honest man, holds his lantern high to reveal to the world his discovery at last of such a man.

When Franklin met Princess Dashkova in Paris in 1781, his fame as a scientist was established, his role as statesman and defender of human rights against tyranny unquestioned, and his work of self-presentation well delineated. The son of a tallow chandler, the printer from Philadelphia, the colonial agent, had become the embodiment of the European Enlightenment and the liberator of his nation, Diogenes' man (*vir*). Dashkova, as a woman and highborn aristocrat, could not transcend in the same way the conditions of her birth, which were the privileges and duties of her rank. The role she had played early in her life as a participant in the coup that overthrew Peter III and brought Catherine II to the throne had led to frequent estrangements from Catherine, periods when she sometimes traveled abroad with her two children. It was during these sojourns abroad that Dashkova was able to enjoy more fully her identity as an intellectual woman. Indeed, although in 1781 Dashkova had yet to achieve much of what we remember her for, she arrived in Paris with a reputation for intellectual achievement, as "a woman of uncommon good parts & great strength of understanding."[47] In this respect Paris was the fitting setting of this meeting between Franklin and Dashkova, the patriot and the princess, for they both belonged to the Enlightenment world whose center was Paris and whose language was that of Montesquieu and Voltaire. And it is as the head of an intellectual society of scholars that the princess would find at last a role at Catherine's court. Of the meeting itself we have only the slightest traces. Dashkova wrote from the Hôtel de la Chine to invite Franklin for the "next Saturday Evening the third of February if convenient to him."[48] Franklin confirmed the meeting in a brief account he wrote in a reply to Georgiana Shipley's letter of introduction: "I recd also your very kind Letter by a Made. Sherbinin[49] with whom and the Princess her Mother I am much pleased; tho' I have not seen them so often as I wish'd, living as I do out of Paris."[50] The princess left Paris in March with her children and traveled slowly back to Russia. There, on January 24, 1783, Catherine II appointed her director of the Imperial Russian Academy of Sciences and Arts in St. Petersburg, a position she held for eleven years. When Franklin wrote to congratulate Dashkova, it was in terms that must have pleased her well: "It gave me great Pleasure to hear, that your magnanimous Empress had plac'd you at the Head of your Academy of Sciences. It was doing Honour to Learning."[51] For her part Dashkova "considered [Franklin] to be a very superior man who combined profound erudition with simplicity of dress and manner, whose modesty was unaffected, and who had great indulgence for other people."[52] In 1789 the American Philosophical Society elected Princess Dashkova to membership, and that same year the Academy of Sciences extended membership to Benjamin Franklin. These reciprocal offers of membership were a final sign of their mutual regard.

Notes

1. *The Spectator* was a daily periodical (except Sundays) founded in 1711 by Joseph Addison and Richard Steele, its principal contributors. The paper, prized for its style and ethics, was comprised of single essays concerned mainly with manners and morals. After a run of almost two years, and a brief revival in 1714 by Addison, the papers were published in book form and sold throughout the century.

2. Samuel Johnson, "Preface to Addison," in *Prefaces Biographical and Critical to the Works of the English Poets* (London, 1779–81), vol. 5, p. 158.

3. Benjamin Franklin, *The Papers of Benjamin Franklin,* eds. Leonard W. Labaree et al. (New Haven, 1959–), vol. 1, pp. 8–45. Franklin experienced an "exquisite Pleasure" on the success of these anonymous essays (See

Benjamin Franklin, *The Autobiography of Benjamin Franklin,* 2nd ed., eds. Leonard W. Labaree, Ralph L. Ketcham, Helen C. Boatfield, and Helene H. Fineman [New Haven, 2003], pp. 67–68).

4. Antoine Arnauld and Pierre Nicole, Jansenist theologians, were the authors of *La Logique, ou l'Art de penser* (1662), known as the *Port-Royal Logic,* an influential handbook. Franklin's copy of the 1717 edition of John Ozell's translation, *Logic, or the Art of Thinking,* is in the Library Company of Philadelphia. See Franklin, *Papers,* vol. 1, p. 58.

5. Xenophon's *Memorabilia,* a compilation of his recollections of Socrates, describes and illustrates the philosopher's method in conversation.

6. Anthony Ashley Cooper, third earl of Shaftesbury, was the author of *Characteristics of Men, Manners, Opinions, Times* (1711); Anthony Collins' best-known work was *Discourse of Free-Thinking* (1713).

7. Franklin, *Autobiography,* pp. 69–71.

8. Franklin, *Autobiography,* p. 80.

9. Franklin, *Autobiography,* p. 106.

10. Franklin's "Plan of Conduct [1726]," in Franklin, *Papers,* vol. 1, pp. 99–100.

11. Franklin, *Autobiography,* p. 146.

12. Franklin, *Autobiography,* p. 200.

13. Franklin, *Autobiography,* p. 117.

14. Franklin, *Autobiography,* p. 130.

15. Franklin, *Papers,* vol. 7, pp. 109–11.

16. These and similar observations occupied Franklin's thoughts every time he crossed the Atlantic. In 1785 during his eighth and final crossing, on his way back to America from France, Franklin elaborated these ideas in a paper he entitled "Maritime Observations" (See Franklin, *Autobiography,* pp. 255–57).

17. Franklin to an unidentified correspondent, London, November 28, 1768 (Franklin, *Papers,* vol. 15, pp. 272–73).

18. Franklin, *Papers,* vol. 4, pp. 125–28.

19. Franklin, *Papers,* vol. 4, pp. 19–20.

20. Franklin, *Papers,* vol. 4, pp. 302–10, 315–17, 466.

21. William Strahan, the London printer and publisher, rushed a copy of Franklin's testimony to David Hall, the Philadelphia printer, saying: "To this very Examination, more than to any thing else, you are indebted to the *speedy* and *total* Repeal of this odious Law" (Franklin, *Papers,* vol. 13, p. 125).

22. Franklin's testimony (Franklin, *Papers,* vol. 13, pp. 124–62) was published immediately in London and then in America; a French translation appeared in 1768. For a description of these publications, see Durand Echeverria and Everett C. Wilkie, Jr., *The French Image of America: A Chronological and Subject Bibliography of French Books Printed before 1816 Relating to the British North American Colonies and the United States* (Metuchen, N.J., 1994), vol. 1, p. 245.

23. Franklin, *Papers,* vol. 19, n. 294.

24. Echeverria and Wilkie, *The French Image of America,* vol. 1, pp. 215–52.

25. Franklin to Samuel Cooper, London, April, 14 1770 (Franklin, *Papers,* vol. 17, p. 123).

26. Franklin, *Papers,* vol. 20, pp. xxxiv, 423.

27. Jacques Barbeu-Dubourg, "Préface du Traducteur," in Franklin, *Papers,* vol. 20, pp. 423–32.

28. Barbeu-Dubourg to Franklin, May 8, 1768, and April 16 [17], 1774 (Franklin, *Papers,* vol. 15, pp. 112–13; vol. 20, p. xvii; vol. 21, pp. 193–94). See also Echeverria and Wilkie, *The French Image of America,* vol. 1, p. 283. The quatrain translates: "He stole fire from the heavens, he makes the arts flourish in the wilderness. America has placed him first among her wise men, Greece would have ranked him with her gods."

29. Barbeu-Dubourg to Franklin, December 29, 1773 (Franklin, *Papers,* vol. 20, p. 520).

30. Franklin to Henry Bouquet, September 30, 1764 (Franklin, *Papers,* vol. 11, p. 367).

31. Franklin to Polly Stevenson, September 14, 1767, January 12, 1777 (Franklin, *Papers,* vol. 14, pp. 254–55; vol. 23, pp. 155–56).

32. In 1777, not long after Franklin's arrival in Paris, a new translation of "The Way to Wealth" (*La Science*

du Bonhomme Richard, ou Moyen facile de payer les impôts) was published in French. The translation, by Antoine-François Quétant, superseded the version Barbeu-Dubourg had published in his 1773 edition of Franklin's works. Quétant's very popular edition, copies of which Franklin gave as gifts, also included translations of Franklin's testimony before the House of Commons in 1766, and the Pennsylvania constitution of 1776. See Echeverria and Wilkie, *The French Image of America,* vol. 1, pp. 344–45; and also letters to Franklin about Quétant's edition in Franklin, *Papers,* vol. 23, p. 587; vol. 24, p. 115; vol. 25, pp. 62–63, 158, 380–81.

33. Durand Echeverria, *Mirage in the West: A History of the French Image of American Society to 1815* (Princeton, 1957), pp. 48–49. Echeverria's book provides a good history of the evolving image of America in Europe; chapter two describes in detail the enthusiasm for Franklin during his mission to France.

34. Franklin, *Papers,* vol. 26, n. 670.

35. Quoted in Franklin, *Papers,* vol. 25, p. lxiii n. 673; vol. 26, p. lxx n. 362

36. Antoine Court de Gébelin to Franklin, June 29, 1778; Joseph-Jerôme le Français de Lalande to Franklin, September 30, 1778; M. Bevos to Franklin, December 28, 1778 (Franklin, *Papers,* vol. 26, pp. 697–98; vol. 27, n. 479; vol. 28, pp. 286–88).

37. See for example an account of the report that he and Jean-Baptiste Le Roy submitted to the French Academy of Sciences on lightning rods for the Strasbourg Cathedral (Franklin, *Papers,* vol. 32, pp. 373–76).

38. Franklin to Giambatista Beccaria, November 19, 1779 (Franklin, *Papers,* vol. 31, pp. 128–29). Beccaria responded enthusiastically to his old friend by sending him a nine-page bibliography of his scientific papers, and later several papers as well (Franklin, *Papers,* vol. 31, pp. 246–47; vol. 34, pp. 380–81).

39. Franklin, *Papers,* vol. 32, pp. 341–39; vol. 34, pp. 120–23, 353, 521–22; vol. 35, pp. 544–46, 548–50; vol. 36, pp. 220–21; vol. 37, pp. 211–12, 504–12.

40. Franklin to Sir Joseph Banks, July 27, 1783, in Benjamin Franklin, *Writings,* ed. J. A. Leo Lemay (New York, 1987), p. 1074.

41. Franklin to Banks, August 30, 1783 (Franklin, *Writings,* p. 1074).

42. Franklin's reply, was reported in Maurice Tourneux, *Correspondance littéraire, philosophique et critique par Grimm, Diderot, Raynal, Meister, etc. revue sur les textes originaux comprenant outre ce qui a été publié à diverses époques les fragments supprimés en 1813 par la censure les parties inédites conservées à la Bibliothèque Ducale de Gotha et l'Arsenal à Paris* (Paris, 1877–82), vol. 13, p. 349.

43. Franklin to Sarah Bache, June 3, 1779 (Franklin, *Papers,* vol. 29, p. 613).

44. Franklin to Thomas Digges, June 25 [1780] (Franklin, *Papers,* vol. 32, pp. 590–92). The miniature was painted by François Dumont after the oil painting by Joseph-Siffrède Duplessis, which Franklin also selected as the model for the miniature he gave to a young friend as a wedding present. See the letters to Franklin from Georgiana Shipley (June 6, 1779) and William Alexander (September 8, 1779) in Franklin, *Papers* (vol. 29, p. 635; vol. 30, pp. 313–16). The miniatures are reproduced in Franklin's *Papers* (vol. 30, facing p. 314; vol. 31, frontispiece).

45. This is a miniature version of a Cathelin's 1779 engraving after a painting by Anne-Rosalie Filleul, which is reproduced in Franklin's *Papers* (vol. 29, p. xxxiii). At the base of the platform are the scattered symbols of liberty associated with Franklin: a broken yoke, an eagle above a cloud of flames and lightning bolts, and a map of North America. Behind Diogenes is a liberty cap atop a staff, the attributes of Libertas, goddess of liberty. In the sky above, a dove has broken free from its tether, and in the background, an open book lies on the rim of Diogenes' tub.

46. The English actually reads "Diogenes has found him alive," but I take *vivum* to be an error for *virum,* "Diogenes has found the man," the text of the commonplace associated with Diogenes.

47. Georgiana Shipley to Franklin, January 6, 1781 (Franklin, *Papers,* vol. 34, p. 253).

48. Dashkova to Franklin, January 30, 1781 (Franklin, *Papers,* vol. 34, p. 327).

49. Anastasiia Shcherbinina is Dashkova's daughter.

50. Franklin to Georgiana Shipley, [after Feb. 3, 1781] (Franklin, *Papers,* vol. 34, p. 347).

51. Franklin's draft letter to "Madame la Princesse Dashkaw" (May 7, 1788) is at the Library of Congress, Washington, D.C.

52. E. R. Dashkova, *The Memoirs of Princess Dashkova,* trans. and ed. Kyril Fitzlyon (Durham, 1995), p. 228.

Documentary Essays

Books Make the Woman:

Princess Dashkova's Moscow Library

ALEXANDER WORONZOFF–DASHKOFF

*P*rincess Ekaterina Romanovna Dashkova devoted much of her life to the printed word. She wrote poetry, plays, essays, travelogues, and an autobiography; edited a dictionary and academic journals; published fiction, nonfiction, and scholarly works; and collected the important books of her time. According to contemporary accounts, her library at the end of the eighteenth century was one of the great private collections in Russia; it included volumes in French, Russian, English, Italian, and German. For a long time it was thought that virtually no record had remained of the titles in her holding, but in 1993, while working at the Alupka Museum in the Crimea, I came across an extensive catalogue of her Moscow library.[1]

This discovery has made it possible to define more precisely Dashkova's main areas of interest, to isolate the works that shaped her own literary and scholarly output, and to establish sources that created patterns of influence for Dashkova and her contemporaries.

The catalogue is in fact an inventory of Dashkova's collection; it was compiled when the books were moved after her death in 1810. There is no consistency in the way the volumes are recorded: often, sets are not listed together and the order is neither alphabetical nor chronological. Entries are written in the same, unknown hand; titles and authors are abbreviated or omitted entirely, making precise identification impossible.[2] Also, it is difficult to determine with any certainty how many of these books Dashkova actually read or how she related to them. As a bibliophile, she might have considered buying any new publication worthy of note, whether it appealed to her personal interests or not. Her employment of Ivan I. Shuvalov, an enthusiastic patron of learning and the arts, as a purchasing agent for "all the literary novelties," and her eager acquisition of a large section of her late uncle Mikhail I. Vorontsov's library, do not suggest careful discrimination.[3] Later, as director of two Academies—the Imperial Academy of Sciences and Arts in St. Petersburg and the Russian Academy—Dashkova would certainly expect to receive copies of their publications and might well attract random presentation copies. Nevertheless, by comparing titles in the catalogue with Dashkova's memoirs and other writings, it is possible to match some of the holdings of her library to her specific areas of interest and professional activity.

In the memoirs, Dashkova informs us that at an early age she threw herself into reading: "Bayle, Montesquieu, Voltaire, and Boileau were my favourite authors."[4] The *Catalogue* supports Dashkova's statement (*Extrait du Dictionnaire de Bayle*; Montesquieu's *De l'Esprit des lois*; *Oeuvres de Boileau*), but, judging by the number of works, Voltaire was most influential (*Oeuvres de Voltaire* [66 vols.]; *Voltaire, Romans*; *Voltaire*; *Pièces de Voltaire*; *La Vie de Voltaire*). Somewhat unexpectedly, Voltaire's dominance in the area of French literature and thought is rivaled only by Jean-Jacques Rousseau (*Emile ou De l'éducation*; *Esprit et maximes*; *Rousseau*; *Pensée de Rousseau*; *Julie ou La Nouvelle Héloïse*), whom Dashkova openly disliked. Her public disdain for the man and his ideas grew even more pronounced after the French Revolution. In the memoirs she admits that a certain aspect of Rousseau's sophism appealed to her as a young adult, but she then goes on to agree with Empress Catherine II that he is a dangerous thinker.[5] Martha Wilmot, an Irish friend and guest of Dashkova's while she was writing the memoirs, wrote: "The P.[rincess] never saw Rousseau. She had too much contempt for him."[6] It is interesting then that so many of Rousseau's writings are recorded in the catalogue; indeed, it seems to suggest that the French author exerted a fair amount of influence on Dashkova's formative years.

Dashkova recalls rereading Claude Adrien Helvétius's *De l'esprit* in 1757. As is often the case in the memoirs, her memory seems to fail her when it comes to chronological accuracy, for *De l'esprit* was published in 1758.[7] However, she indisputably read the French philosopher, since a copy of *De l'esprit,* annotated by Dashkova, was discovered in the library of the Academy of Sciences.[8] She also contributed her translations of Helvétius and Voltaire to the journal *Nevinnoe uprazhnenie* [Innocent Exercise]. *De l'esprit* and other works by Helvétius appear in the catalogue, as do most of the journals, serial publications, and scholarly editions associated with her name. As director of the Academy of Sciences she published and edited the *Sobesednik liubitelei rossiiskogo slova* [Companion for Connoisseurs of the Russian Word] (15 vols.) and *Novye ezhemesiachnye sochineniia* [New Monthly Essays] (2 sets, 20 and 28 vols.); her articles were printed in *Opyt trudov Vol'nogo Rossiiskogo Sobraniia* [Works of the Free Russian Assembly] (6 vols.) and *Drug prosveshcheniia* [Friend of the Enlightenment] (12 vols.); as a member of the Free Economic Society, she subscribed to their *Trudy* [Works] (7 vols.); she collaborated on the anthology *Rossiiskii featr* [Russian Theater] (44 vols.), in which her play *Toisiokov* (Mr. This-and-That) appeared. Indeed, these and the many other serials listed in the catalogue (*Journal encyclopédique* [22 vols.]; *Ekonomicheskii magazin* [The Economic Magazine] [44 vols.]; *The Tatler* [4 vols.]; *The Bee* [3 vols.]) attest to the diversity and international scope of her interests.

At the Russian Academy, Dashkova spent years supervising and assisting in the compilation of the *Slovar' Rossiiskoi Akademii* [Dictionary of the Academy of the Russian Language] (6 vols.). Praised by the poets Nikolai M. Karamzin and Aleksandr S. Pushkin, among others, the etymological dictionary provided some order and standardization to eighteenth-century Russian usage. Also listed in the catalogue is the *Sravnitel'nyi slovar' vsekh iazykov* [Comparative Dictionary of All Languages] (4 vols.), a project Catherine II advanced over Dashkova's own dictionary. Catherine's decision to favor the other work contributed to competition and ill feelings between the two women. In the memoirs Dashkova recalls with bitterness that "useless and imperfect as this peculiar work was, it was pronounced to be an admirable dictionary and caused me considerable annoyance."[9]

On the whole, Dashkova's library contains an impressive assortment of reference works (*Slovar' iuridicheskii* [Law Dictionary]; *Tserkovnyi slovar'* [Ecclesiastical Dictionary]; Samuel Johnson's *Dictionary*). She possibly knew James Boswell and others of Samuel Johnson's circle; in fact, many of the works given in the catalogue were written by individuals she had encountered during her travels abroad. In Edinburgh, she met William Robertson (*History of the Reign of the Emperor Charles V* and *History of Scotland*), Hugh Blair (*Lecture on Rhetoric and Belles Lettres*) and Adam Smith (*An Inquiry into the Nature and Causes of the Wealth of Nations*): all of whom Dashkova claimed "came twice a week to spend the day with me."[10] Dashkova recommended Smith's book to her brother Aleksandr R. Vorontsov, then president of the College of Commerce. She corresponded with David Garrick (*Garrik. Akter* [Garrick. The Actor]), attended gatherings held by the Bluestockings (*The Work of Miss Hannah More*), and claimed to have spent long hours with Denis Diderot (*Oeuvres morales*) and the Abbé Raynal (*Mémoires de l'Europe, Atlas pour l'histoire des établissements*) in Paris. In Rome she was granted an audience with Pope Pius VI (*Le reflessioni di Papa Pio VI*) and discussed with him the formation of a museum in the Vatican.

The philosophical, literary, and historical texts in the catalogue sketch a map to Dashkova's intellectual pursuits and to the wider panorama of eighteenth-century Russian culture. Dashkova collected Longinus, Confucius, Etienne Bonnot de Condillac, Nicolas Malebranche, Francis Bacon, David Hume, *Bogoslovskie razsuzhdeniia* [Theological Discourse, by an unidentified author] and Emanuel Swedenborg's *Les Marveilles du ciel et de l'enfer*. Her library included French literature (Molière, Jean François Marmontel, Jean François de Laharpe, Montaigne, La Fontaine, Pierre de Beaumarchais, duc de La Rochefoucauld), classical and Italian literature (Homer, Virgil, Catullus, Petrarch, Boccaccio, Dante, Ariosto, Goldoni), and other works such as *Mélange de litterature orientale*. Many volumes appear in French translation (Johann Wolfgang von Goethe's *Werther*, Friedrich Schiller's *La Pucelle d'Orléans*; Alexander Pope's *La Dunciade*), and despite a limited command of English, as revealed in her notes and letters to Martha Wilmot, Dashkova's library contained an extensive collection of verse and prose in the original language (Johnson's and Steven's *Shakespeare*, Chaucer, Milton, Donne, Dryden, Swift, Addison, Richardson, Smollett, Sterne, *The Poems of Ossian*). While many Russian authors were removed from Dashkova's collection earlier, those that remain bear directly on her activities in the field of literature (*Sochineniia Kniazhnina* [The Works of Kniazhin]; Radishchev's *Zhitie Fiodora Vasil'evicha Ushakova* [The Life of Fedor Vasil'evich Ushakov]). Her publication at the Academy of Sciences's expense of Iakov B. Kniazhin's controversial *Vadim Novgorodskii* [Vadim of Novgorod] angered Catherine and may have contributed to Dashkova's early "retirement." In the memoirs Dashkova claims that having read Aleksandr N. Radishchev's work, she had correctly predicted the unfortunate consequences that befell her brother on account of his protégé's political views. Dashkova therefore was not surprised when she received a letter from her brother informing her that Radishchev had published his *Puteshestvie iz Peterburga v Moskvu* [*Journey from St. Petersburg to Moscow*] (1790), a "tocsin to revolution," as she referred to it.[11] Written in the form of a travelogue, it was a thinly disguised condemnation of serfdom, tyranny, and other government abuses. Its allegedly revolutionary content outraged Catherine II, who had Radishchev apprehended, interrogated, and exiled. Officially, Vorontsov requested a leave of absence, but in reality he was forced into retirement and subsequently resigned his governmental post. He continued to support Radishchev and his family financially during their exile in Siberia, supplying them with money and other necessities.

The catalogue offers other possible pointers for establishing the authors and works that shaped and defined Dashkova's own literary output. For instance, the memoirs are the product of both a classical education and a later introduction to the sentimental tradition, which engaged a conscious effort to induce and analyze emotion in art, literature, and one's own life. In this regard, foreign, mostly French, literature served as an important model: from Rousseau's confessional mode to the numerous official memoirs written by statesmen, ministers, generals, marshals and diplomats (Sully,

Maurepas, Bonneval, Noailles, Baron de Tott, Bassompierre). Also of interest was *Mémoires du Ct. de St. Germain*, especially since Vorontsov knew the author personally. Although none of these autobiographies were written by women, the catalogue lists two volumes of the poet Antoinette Deshoulières's works, and Mme de Sévigné's *Lettres de Sévigné* (9 vols.) must have been a source of great inspiration. Like Dashkova, Mme de Sévigné was raised by an uncle, married early, had two children and was widowed at a young age. In view of Dashkova's complex and ultimately tragic relationship with her children, she must have been deeply touched by Mme de Sévigné's intelligent and highly dramatic correspondence with her daughter.[12]

The makeup and focus of Dashkova's library mirror and elaborate on the three primary areas of interest in the memoirs: the coup d'état of 1762; travel in the West; and her duties at the Academy of Sciences. Shortly prior to the events that brought Catherine II to the throne Dashkova was "reading everything that had ever been written on the subject of revolutions [i.e. political change] whenever they happened to occur."[13] The catalogue attests to the polemical nature of Dashkova's memoirs and to her lifelong interest in the nature and consequences of sociopolitical upheaval and change (Vertot's *Révolutions Romaines*; *Révolution de Portugal*; *History of Revolution in Sweden*). She was understandably fascinated by the reign of Catherine the Great, particularly when her own participation was highlighted (*Instruction de Catherine II; Instruction the Commis. a New Code Russ. Empire, Nakaz Ekateriny II* [The Instructions of Catherine II]; *Oeuvres posthumes de Rhulière*). Dashkova read Rulhière's history in manuscript form and expressed her dissatisfaction with it in the memoirs. She was also deeply interested in general Russian history (Nicolas-Gabriel Le Clerc's *Histoire de Russie*; Pierre-Charles Levesque's *Histoire de Russie*; *L' Histoire de Russie de Lomonosow*). A major area of inquiry was Peter the Great's influence on the development of Russia when compared to the enlightened rule of Catherine (*Journal de Pierre le Grand*; *Mémoires du règne de Pierre I*; *Istoriia Petra Velikogo* [History of Peter the Great]; *Zapiski Petra Velikogo* [Memoirs of Peter the Great]). For instance, Dashkova discusses Peter the Great in her memoirs and refers to the Russian tsar as "quick-tempered, brutal, and despotic."[14] Interestingly, in the memoirs Dashkova passes over the American Revolution, yet her concern with colonial unrest in the New World is reflected by the large number of books on the subject (*Histoire des troubles de l'Amérique Anglaise*; *Abrégé des révolutions de l' Amérique Anglaise*; *Political Pieces by Franklin*).

Dashkova was an enlightened traveler, deriving for her children and herself maximum educational benefit from the places she visited (*Description de la gallerie de Florence*; *Commerce de Tarente*; *Observations sur Londres*; *A History of Edinburgh*). She describes in the memoirs how every morning in Brussels "Dr. Burtin and I went off botanizing together in the neighbourhood of the city where I found several plants we had never seen at home."[15] Apparently, such excursions were a regular feature of her travels (*Histoire des plantes aux environs de Paris*). Moreover, travel offered her the opportunity to study at close quarters and evaluate for herself various political, legislative, and legal structures (*Edit de la République de Genève*; *Istoria della Republica di Venezia*; *Legislative Rights*; *Institution of the Law in Scotland*; *Political Survey of Ireland*). As a member of Nikita Panin's circle, she strove to broaden her understanding of constitutional systems and limited monarchies (*Constitution de l'Angleterre*; *La Constitution Française*; *Difference Between Absolute and Limited Monarchy*). In defiance of her falling-out with the Catherine and exile from court, Dashkova hoped to continue contributing to the course of political change in Russia, either through direct or indirect statecraft (*Kak dogovarivat'sia s gosudariami* [How to Negotiate with Sovereigns]; *Opera di Machiavelli*). Unfortunately, the latter was not her strong suit.

As head of the Academy of Sciences and the Russian Academy (*Mémoires de l'Académie des Sciences*; *Commentarii Academiae*; *Uchenye razsuzhdeniia Akademii Nauk* [Scholarly Discourses of the Academy of Sciences]; *Akademicheskie izvestiia* [*Academic News*]), Dashkova made her mark in letters and science at some distance from Catherine and the throne. Through her travel and association with a number of academies, learned societies, and institutions of higher learning (*Mémoires de l'Académie des Sciences de Berlin*; *Transactions of the American Society*; *Histoire de la Sorbonne*), she served as an important intermediary in the dissemination of the ideas of European Enlightenment at home and the introduction of Russian culture in the West. The catalogue of her books includes works on poetics, linguistics, geography, geology, botany, zoology, and archaeology, to name but a few. It also contains works from the specialized fields of her friends and acquaintances such as Mikhail V. Lomonosov and Benjamin Franklin (*Histoire de l'éléctricité*; *Essai sur l'éléctricité; Theoria Electricitatis*) and the writings of her adversaries (Peter Simon Pallas's *Puteshestvie* [Journey]). She collected a great number of works by Leonhard Euler (*Recherches et calculs d'Euler*, *Calculus integralis*; *Analytica*), whom she considered to be "the greatest geometrician and mathematician of his age."[16] In addition to scholarship and research, Dashkova felt that the Academy of Sciences should play an important role in the expansion and improvement of secondary and higher education. She was interested both in theoretical questions based on foreign models (*Traité sur l'éducation* [*des filles*, Fénelon?]; *Hume on Education*; *Lokka vospitanie detei* [Locke's Education of Children]) and in practical application (*Pravila dlia narodnykh uchilishch* [Regulations for the People's Schools]; *Opyt narodnogo vospitaniia* [Essay on the People's Upbringing]; *Ustav vospitaniia blagorodnykh devits* [Statute on the Education of Young Noblewomen]).

Apart from strictly professional considerations, Dashkova's pedagogical and scholarly activities extended to her private life and to the guidance and tutoring of her children. Military science played an important part in her son Pavel's education (Mallet's *Fortification novelle*, *Elementi di archithethura militare*, *Kniga o voennom iskusstve* [Book on Military Science]) and books on the proper upbringing and conduct of women are well represented in the library (*Discours sur l'éducation des dames*; *Education de la noblesse française*; *Cour d'éducation des demoiselles*; *Dolzhnosti zhenskogo pola* [The Duties of Women]). Finally, the catalogue lists monographs, treatises, and instructional tracts on subjects that Dashkova felt were within her special areas of expertise: architecture (*Principes de l'architecture*; Vilroy's *Dictionnaire d'architecture*; *Views of Public Buildings*; *Kratkoe rukovodstvo k grazhd. Arkhitekture* [A Short Guide to Public Architecture]); art and music (Félibien's *Entretiens sur les vies des peintres*; *Essay on Painting*; *The Present State of Music*); agriculture (*Elemens d'agriculture*; Young's *Annals of Agriculture*; *O zemledelii* [On Agriculture]), especially when it concerned the estates in her region (*Opisanie kaluzhskogo namestnichestva* [Description of the Kaluga Region]); gardening (*Théorie des jardins*; *Dissertation on Oriental Gardening*); health and medicine (Grimant's *Memoires sur la nutrition*; *Observations sur le manuel des accouchemens* [sic]; *History of Drugs*; *Nyneshnii sposob privyvat ospu* [Current Method of Smallpox Inoculation]).

The fate of Dashkova's Moscow library remains uncertain and it is unclear whether any of its volumes survived the fire of 1812. It is, however, possible that Mikhail S. Vorontsov, Dashkova's nephew and heir, sent the books to his residence in Odessa where they were integrated into his library. In 1898, the Vorontsov family presented this collection to Odessa University, where it is located to this day. Therefore, further study of the catalogue will assist greatly in the possible identification and recovery of additional volumes and will continue to suggest new elements in the education, literary works, scholarship, and intellectual pursuits of a truly remarkable eighteenth-century woman.

Notes

1. The manuscript entitled *Catalogue* in French consists of fifty-two, unpaginated, gold-trimmed sheets, 32 cm × 20.5 cm, bound by a variegated silk thread; watermark 'Cantscharof 1810, Propatria' (woman with sceptre, lion with crown). It is located in the M. S. Vorontsov Memorial Library, Alupka Museum in the Crimea.

2. Titles of works and authors are given in parentheses throughout the essay as they appear in the *Catalogue*; that is, often incompletely. To avoid confusion, authors' first names are at times provided.

3. E. R. Dashkova, *The Memoirs of Princess Dashkova,* tr. and ed. Kyril Fitzlyon (Durham, 1995), p. 35.

4. Dashkova, *Memoirs*, p. 33.

5. Dashkova, *Memoirs*, p. 235.

6. Martha and Catherine Wilmot, *The Russian Journals of Martha and Catherine Wilmot*. eds. The Marchioness of Londonberry and H. Montgomery Hyde (London, 1935), p. 142.

7. Dashkova, *Memoirs*, p. 34.

8. f. 5154, no. 2015.

9. Dashkova, *Memoirs*, p. 216.

10. Dashkova, *Memoirs*, p. 147.

11. Dashkova, *Memoirs*, p. 237.

12. Dashkova disinherited her daughter and was so angered by her son's conduct that she did not attend his funeral.

13. Dashkova, *Memoirs*, p. 64.

14. Dashkova, *Memoirs*, p. 181.

15. Dashkova, *Memoirs*, p. 155.

16. Dashkova, *Memoirs*, p. 205.

A Tumultuous Life:

Princess Dashkova and Her Relationships with Others

SVETLANA ROMANOVNA DOLGOVA

*P*rincess Ekaterina Romanovna Dashkova (born Vorontsova) was a woman of many contradictions. Her life has long attracted the attention of historians both in Russia and abroad, yet she remains elusive because much of the relevant archival material is unpublished, and what has been published is not widely available. In addition, during the last century, scholarly work on Dashkova was inhibited because her role as a court lady was problematic in the eyes of certain Soviet ideologues. Today, however, her contribution to eighteenth-century Russia is widely acknowledged even though aspects of her life remain controversial.[1]

Princess Dashkova and Catherine the Great

Dashkova was one of the strongest supporters of Catherine the Great; she even helped in the coup that brought Catherine to power in 1762. Over the years their friendship changed. In the beginning, prior to Catherine's ascension to the throne, the two women had a deep emotional connection to one another. Later, when court life and politics intervened, their friendship turned to cool detachment relieved only by periodic, politically motivated admiration.

The two Ekaterinas (or Catherines),[2] "the Great" and "the Little" as they are sometimes called, met in the winter of 1759, when the Grand Duke Peter, heir to the Russian throne, and his German-born wife, the Grand Duchess Ekaterina Alekseevna, visited Grand Chancellor Mikhail I. Vorontsov on Sadovaya Street in St. Petersburg. The young Ekaterina Vorontsova was Mikhail's niece, and she had lived with him since she was four years old. At the time of this first meeting, Dashkova was only fifteen, half the age of the grand duchess. However, despite her youth, the young girl impressed the future empress with her conversational skills, her outstanding knowledge of progressive literature, and her profound intellect.

On June of 1762, Dashkova actively participated in the coup that overthrew Peter III, Catherine's husband. Catherine's ascendance to the throne brought Dashkova many honors: she was awarded entrance into the Order of St. Catherine and received a large monetary compensation. But Dashkova exaggerated her part in the coup in an attempt both to remain close to the empress and to keep Count Grigorii G. Orlov, Catherine's lover, from the throne. Dashkova's distrust of Orlov was undoubtedly one of the reasons that Catherine's relationship with Dashkova cooled shortly after Catherine became empress. Even foreign ambassadors to the court immediately noticed the change, often describing Dashkova as a conceited and talented woman whose energy was threatening even to the empress. For example, English ambassador John Hobart, the second earl of Buckinghamshire, wrote of Dashkova in a report on June 28, 1763, a year after the coup:

> That lady's arrogant behaviour had in great measure lost the Empress's esteem . . . Her spirit was too great either to try to appease her Mistress or to submit to her disgrace, and she has been suspected ever since of exciting and encouraging those who were disaffected by the present government.[3]

Two years later another English ambassador, Sir George Macartney, the first earl Macartney, reported that Dashkova found a cold and unwelcoming reception at the Russian court. He added, however:

> She is a woman of uncommon strength of mind, bold beyond the most manly courage and of a spirit capable of undertaking impossibilities to gratify any predominant passion—a character

highly dangerous in a country like this, especially when joined to an engaging behaviour and a beautiful person.[4]

Catherine's jealousy of Dashkova may also have been sparked when she learned that Ivan I. Shuvalov, the former lover of the late Empress Elizabeth and a very powerful figure of the time, had written to Voltaire shortly after the coup, "a nineteen-year-old woman has changed the government of this country." Catherine immediately asked Stanislav A. Poniatovskii, one of her former lovers, to "please teach this great author better!"[5]

In her letter to Poniatovskii, Catherine made it very clear that she herself was the only person who knew everyone and everything about the plot that led to the coup and, therefore, only she herself deserved to receive credit for it. She claimed that Dashkova had too many flaws to be considered a serious participant even though she was personally acquainted with some of the key players. Catherine charged: "On account of her family connections and her age—only nineteen—she did not stand in good repute; she inspired confidence in no one." Catherine further wrote: "All of the conspirators had been in touch with me for six months before [Dashkova] even knew their names . . . during the last four weeks she was told as little as possible."[6] Elsewhere Catherine also noted: "She is very clever, but besides her great vanity she has a muddle-headed character, and our leaders did not like her. Only thoughtless people put her in possession of what they knew and this consisted only of small details."[7] It is difficult to know the motivation behind such statements, as the empress, like all sovereigns, had to protect her own interests; but it is clear that Catherine's feelings about Dashkova were ambivalent at best.

In addition to Catherine's dislike of Dashkova's ambition, she did not like the Vorontsovs, Dashkova's family; she considered them snobs who were opposed to her rule. For example, Dashkova's uncle Mikhail (grand chancellor under both Elizabeth I and Peter III, and hence one of the most powerful men in Russia) did not rush to take the oath of allegiance to Catherine when she ascended to the throne. He waited until after the death of Peter III, which occurred about a week after the coup. Also, Catherine knew that Peter had been eager to legalize his relationship with Elizabeth Vorontsova, Dashkova's sister. The fact that Peter's mistress was a Vorontsov was probably another reason for Catherine's coolness towards Dashkova.

Contrary to expectations, however, Catherine II did not do away with her husband's mistress when she came into power. Instead she immediately sent Elizabeth back to her father's house in St. Petersburg, where Dashkova found her surrounded by soldiers. Dashkova asked an officer on duty to reduce the number of guards, precipitating a harsh reprimand from the empress—the first scolding Dashkova received from her friend. Deeply saddened, Dashkova did not reply to Catherine but rather attempted to return the badge and ribbon of the Order of St. Catherine that the empress had given her for her role in the coup. As the nineteenth-century writer Aleksandr I. Gertsen (Herzen) recounts: "'Not so fast, not so fast,' said the empress. 'I had to reprimand you for your impetuosity—you had no right to dismiss the soldiers on your own authority; but I must reward you for your services.'" She then replaced the ribbon around Dashkova's neck.[8] Perhaps at this moment Dashkova recalled the prophetic words of Peter III, who had once said to her: "Remember that it is safer to have to do with honest simpletons like your sister and me than with great intellects who squeeze every drop of juice out of you and then throw you out of the window like the skin of an orange."[9]

Over the years there were numerous rumors that circulated about Dashkova. She was suspected of being involved in an unsuccessful plot to free Ivan IV from captivity in July of 1764,[10] and later, in 1773–74, to have participated in a plot favoring the ascension to the throne of Grand Duke Paul Petrovich, Catherine's minor son.[11] Finally, Dashkova was associated with two authors unpopular with Catherine. The first was Aleksandr N. Radishchev, who wrote a revolutionary book titled *Journey from St. Petersburg to Moscow [Puteshestvie iz Peterburga v Moskvu],* which decried the practice of serfdom in Russia. The second was Iakov B. Kniazhin, whose tragedy *Vadim of Novgorod [Vadim Novgorodskii]* Dashkova published in 1793 when she was director of the Imperial Academy of Sciences and Arts in St. Petersburg. Her association with both men, whether entirely true or not, prompted a severe reaction from Catherine II, who by then was terrified by any hint of rebellion, fearing that the extreme violence and republican ideals of the French Revolution could easily infect Russia.[12]

In summary, though Catherine admired Dashkova's intellect and abilities, it appears that she remained conflicted about her. Imperial secretary Aleksandr V. Khrapovitskii, who recorded many of Catherine's sharp comments about Dashkova in his diary, wrote in April of 1789 that "[the Empress] did justice to Dashkova for having knowledge and skills of many men but was surprised why nobody liked her."[13] In a court rife with petty intrigues and full of nobles competing for imperial favors, such personal comments remain difficult to interpret, as they were rarely free of political concerns.[14]

Dashkova and the Imperial Academy of Sciences and Arts

Catherine II appreciated the keen intelligence, the unrestrained energy, and the considerable renown of her most visible lady-in-waiting, and, when she appointed Dashkova as the director of the Imperial Academy of Sciences and Arts, she must have made the appointment in the interest of the public good. Indeed, her choice would seem to indicate that at this moment in her reign the importance of natural philosophy prevailed over the petty disagreements and unpleasant feelings that had alienated her from her friend.[15]

Dashkova was hesitant to accept the directorship of Russia's most esteemed learned society. But once she took over the position, she quickly proved her capabilities as an active, well-organized leader with a genuine devotion to science. The particular tenor of Dashkova's reign at the Academy became apparent immediately. As she later wrote in her memoirs, she resolved to "squander nothing and stop all misappropriations," being "determined not to enrich myself at the Academy's expense."[16]

One of Dashkova's important innovations was to open lectures by academicians to the general public. These lectures lasted from 1785 to 1802. In a report to Catherine II on April 8, 1784, Dashkova wrote: "Giving lectures in the Russian language not only for the students and pupils but for all lay people seems to me useful because in this manner science will enter into our everyday language, furthering the cause of education."[17] Further, in her memoirs, Dashkova said: "I often attended these lectures myself and had the satisfaction of seeing children of impoverished Russian gentry and young subalterns in the Guards derive benefit from them."[18]

Dashkova also energized the Academy's publishing activities. On her initiative, for instance, the institution published the first anthology of the works of the famous scientist and poet, Mikhail V. Lomonosov.[19] Perhaps more significant was Dashkova's desire to publish a dictionary of the Russian language after she became the founding president of the Imperial Academy of the Russian Language. She succeeded in convincing intellectuals of different social standing of its importance and in ushering it through to completion. Among the authors of this dictionary were the empress herself, as well as individuals who by the nineteenth century would come to be known as *raznochinets* (intellectuals of non-noble descent). Among them were such personages as Fedor V. Karzhavin, a follower of the architect Vasilii I. Bazhenov. An educated man who had studied languages, natural history, and philosophy in Russia and France, Karzhavin came to America in the 1770s and helped the revolutionary cause, first serving as a translator and later organizing shipments of military supplies from the Caribbean.[20] Dashkova herself wrote several entries and persevered over a five-year period to realize the dictionary. Unfortunately but not surprisingly, it became another unpleasant experience for Dashkova at the court, which she wrote about with some bitterness in her memoirs:

> The more enlightened members of the public recognized my merits
> and admitted that the foundation of a Russian Academy and the
> amazing speed with which the Russian dictionary—the first we ever
> had—had been compiled were due to my zeal and public spirit. The
> courtiers, on the other hand, found the dictionary inconvenient, based
> as it was on etymological principles.[21]

During her tenure Dashkova developed many friendships with scientists in Moscow, especially with Alexei F. Malinovskii, a historiographer who would later be married to Anna P. Islenieva, a granddaughter of one of Dashkova's cousins. This relationship became especially important to Dashkova in her later years, when she was excluded from court life and isolated at her estate, Troitskoe. At that time Malinovskii kept her informed by sending her the latest publications as well as relevant news from Moscow.[22]

Overall, Dashkova's relationships with academicians were favorable. Still preserved in the Russian State Archive of Ancient Acts are petitions that awarded worthy scientists and allocated the pensions of the academicians. However, we also know of conflicts that arose when Dashkova would not compromise her principles. Such was the case with junior scientific assistant Vasilii F. Zuev.[23] When he was late in turning in travel journals from an expedition, Dashkova expelled him from the Academy over the protest of his scientific advisor, Peter Simon Pallas, another explorer and academician. She wrote that she took the action "with regret but to show others an example."[24]

Dashkova and the Court

Dashkova held strong opinions on many different issues and defended them rigorously with no regard for the rank or position of her opponents at the court. Her viewpoints are evident in the satirical writings she penned for *Companion for Connoisseurs of the Russian Word,* a journal she founded and to which she contributed ten articles. For the most part, her pieces revealed and mocked vices she saw in contemporary high society, such as ignorance, servility, and hypocrisy. Needless to say, her barbs at the nobility gave many of her highly placed contemporaries a reason to feel animosity towards her.

Dashkova was also capable of vengeance. Perhaps the best example—and certainly the most humorous one—entailed a quarrel over the fate of two pigs owned by Aleksandr Naryshkin, a neighbor who happened to be the son of Peter I's cousin. When the pigs wandered onto Dashkova's property near her *dacha* (summer home), the princess ordered them killed. Upon learning of the dispute that ensued, the empress did not defend Dashkova but instead ordered that the issue be taken to court. During the hearings that took place on November 3, 1788, it was explained that Dashkova spotted the two pigs in her yard and ordered her servants to herd them into a horse stable and kill them; the act was carried out immediately. The court ordered Dashkova to pay Naryshkin eighty rubles for having broken the law.[25] After the incident Naryshkin apparently pointed to Dashkova's face, which was red from anger and indignation, and said: "It is still covered with blood after the murder of my pigs." C. F. P. Masson, a French tutor at the Russian court, later claimed that Naryshkin paused significantly after saying "murder of" and before saying "my pigs," to suggest that Dashkova was guilty of another sacrifice—that of Peter III a quarter-century earlier.[26]

This killing of the two pigs was for Dashkova the resolution of an old argument between her and the Naryshkins over a disputed property. It was also perhaps vengeful in that Naryshkin's wife, Anna Nikitichna (born Rumiantseva) did not like Dashkova. Further, there might have been reason for Dashkova to be jealous of her. According to the memoirs of Catherine II's secretary, Khrapovitskii, both women participated in Catherine's coronation; however, there had been "an order to clean up rooms for Anna Nikitichna but to make arrangements so that there would be no rooms for Dashkova." He claimed that the empress told him: "I want to spend time with one woman but not with the other; they are in the middle of an argument over a pitiful piece of land."[27] Such petty intrigues were not uncommon at the court, and while such a statement appears insulting to Dashkova, it probably had no lasting effect on her relationship with Catherine.

Dashkova and Her Family

Since many at the court held ill will for Dashkova, she would have benefited from a supportive family. But the young Ekaterina lost her mother when she was still a baby, and thus she grew up in the house of her uncle, Mikhail I. Vorontsov, along with his daughter, Duchess Anna Mikhailovna. Both girls were goddaughters of the Empress Elizabeth and her nephew and heir, Grand Duke Peter (the future husband of Catherine). Vorontsov spent a considerable amount of money to give his daughter and niece a good upbringing. They spoke four languages, danced and painted beautifully, and learned refined manners. In Dashkova's later opinion, however, nothing was done to develop their minds and educate their hearts. In her memoirs she claims that it was her obsession with reading and philosophizing that made her a serious and hardworking woman.[28]

Ultimately, despite the generosity of her uncle, the competition with her cousin and the unintentional jealousy between them contributed to a feeling of isolation within the family. Perhaps as a reaction to her upbringing, Dashkova developed an uncompromising and original character. After her marriage to Prince Mikhail I. Dashkov, she seems to have mellowed but her husband died very young, when she was only twenty-one. A widow with two children and many large debts, Dashkova responded to her new circumstances by becoming a scrupulous money manager. She distinguished herself from other Russian noblewomen by entering public service—an exception not only for the eighteenth century but for later periods as well.

Within her family Dashkova was respected and was said to resemble her older brother, Aleksandr R. Vorontsov, who according to his contemporaries "was an original and firm man; his voice was feared in the State Council because he did not try to accommodate the empress's favorites."[29] Both were public servants, and after their respective retirements from public life in the early 1790s, the two siblings became quite close. Dashkova is known to have visited Aleksandr at the Vorontsovs' ancestral estate in the village of Andreevskoi in the Vladimir region. Yet the Vorontsov family as a whole, like Catherine II, seems to have treated her with ambivalence. A letter written by her uncle Mikhail to Aleksandr shortly after the coup that brought Catherine to power indicates that he feared her uncompromising character might reflect badly on the family:

> She, as it seems to me, has a spoiled and conceited character, she spends
> her time in an insincere passion for higher reason and science; she wastes
> her time. I'm afraid that she may make the empress mad by her capricious
> and unchecked behavior and speeches. She then could be distanced from
> the court and our family would suffer the public's reproach through her

fall. To be fair, she did play a large role in our merciful monarch's successful coming to the throne and we must give her glory and respect for that.[30]

Dashkova's relationship with her children was equally complex. The princess took special care to give her son, Pavel, the kind of good education that would allow him to excel in military service. She wrote a curriculum for him at Edinburgh University and spent three years in Scotland while he completed his degree. In Russia, Pavel served under Grigorii A. Potiomkin (Potemkin) and at one point was on active duty in the south of Russia. On January 14, 1788, however, Pavel married a merchant's daughter, Anna S. Alfiorova, without informing his mother. The two were of unequal social rank, which greatly upset Dashkova. Despite the rupture this marriage caused, Dashkova continued to pay off his debts quite frequently. Pavel, on his side, later asked both Emperor Paul I and the emperor's son, the future Alexander I, to allow his mother's return from exile. When Pavel died in 1807, their relationship remained unresolved.

Dashkova's daughter Anastasiia also caused her much anguish. She married Andrei Shcherbinin in 1785, but did not have an easy life with him. The Shcherbinins fought frequently and lived separately for long periods of time. Anastasiia lived with her mother and brother throughout their three-year sojourn in Scotland and their travels through Europe. Furthermore, Anastasiia, like her brother, constantly accumulated debts, which Dashkova had to pay. Nonetheless, Dashkova later recalled how shaken her daughter was by the news of her exile: "I had great difficulty in trying to revive my daughter's hopes and courage, while she cried and clung to my knees." [31]

When Anastasiia offended some of Dashkova's friends at Pavel's funeral, the ties between them were apparently severed. For this and perhaps other unknown reasons, Dashkova subsequently denied her daughter any inheritance and forbade her to enter her house. According to a letter found recently in the Russian State Archives, Anastasiia's cousin Mikhail S. Vorontsov was the sole beneficiary of both her mother's and brother's estates. Anastasiia writes in the letter that she could have made peace with her mother if others had given them the opportunity to see each other and also says she regretted not being given a chance to say goodbye to her mother. She also accuses the executor of Dashkova's will of stealing her mother's library.

In sum, in the archival materials discovered thus far, it appears that Dashkova was admired by her family and her compatriots for her strong intellect and abilities, but it is also clear that her complex, impassioned, volatile, and uncompromising character often resulted in troubled relationships.

Notes

1. After translation of this essay from the original Russian by Irina Dubinina, quotations and other information were added by the editor as a way of providing background and context for an English-speaking audience.

2. The Russian version of the name Catherine is Ekaterina; hence Catherine the Great is Ekaterina the Great.

3. Quoted in Anthony Cross, "Contemporary British Responses (1762–1810) to the Personality and Career of Princess Ekaterina Romanovna Dashkova," in *The Semiotics of Russian Cultural History*, eds. Alexander D. Nakhimovsky and Alice Stone Nakhimovsky (Ithaca, N.Y., 1985), p. 42.

4. Quoted in Cross, "Contemporary British Responses," p. 44.

5. Appendix to *Memoirs of Catherine the Great*, trans. Katharine Anthony (New York, 1927), p. 318. A slightly different translation of Catherine's correspondence with Poniatovskii (with some abridgements) appears in *The Memoirs of Catherine the Great*, ed. Dominique Maroger (New York, 1961), pp. 270–77.

6. Appendix to *Memoirs of Catherine the Great*, p. 318.

7. Appendix to *Memoirs of Catherine the Great*, p. 318.

8. Alexander Herzen, "Princess Yekaterina Romanovna Dashkov (1857)," in *My Past and Thoughts: The Memoirs of Alexander Herzen*, trans. Constance Garnett, rev. by Humphrey Higgins (New York, 1968), vol. 4, p. 1614.

9. Herzen, "Princess Yekaterina Romanovna Dashkov," p. 1598.

10. Cross, "Contemporary British Responses," p. 43.

11. S. R. Dolgova, *Kniaginia E.R. Dashkova i sem'ia Malinovskikh* (Moscow, 2002), pp. 169–173; quoting "Svedeniia dlia zhizneopisaniia kniagini Ekateriny Romanovny Dashkovoi," Aleksei F. Malinovskii's manuscript biography of Dashkova.

12. Other accounts of the alienation between Dashkova and Catherine include that of the nineteenth-century academician Ya. K. Grot, who claimed that Catherine was angry at Dashkova for correcting and maybe even making fun of essays she sent into *Sobesednik liubitelei rossiiskogo slova* [Companion for Connoisseurs of the Russian Word], a journal that Dashkova published and edited while at the Academy.

13. A. V. Khrapovitskii, *Dnevnik A. V. Khrapovitskago* (Moscow, 1901), p. 161.

14. For more on Dashkova's relationship with Catherine, see O. I. Eliseeva, "Ekaterina II i E. R. Dashkova: Fenomen zhenskoi druzhby v epokhu prosveschenia" in *E. R. Dashkova i A. S. Pushkin v istorii Rossii*, ed. L.V. Tychinina (Moscow, 2000), pp. 19–33.

15. See Michael Gordin's essay, "Arduous and Delicate Task," in this volume.

16. E. R. Dashkova, *The Memoirs of Princess Dashkova*, trans. and ed. Kyril Fitzlyon (Durham, 1995), p. 206.

17. Archives of the Russian Academy of Sciences, PFA RAS f. 3, op.1, d. 556, L. 215–215 ob.

18. Dashkova, *Memoirs*, p. 208.

19. Lomonosov had died in 1765. He is often called the greatest Russian scientist of the eighteenth century; his wide-ranging interests included physics, chemistry, electricity, the manufacture of glass and porcelain, and Russian language and literature. See Boris N. Menshutkin, *Russia's Lomonosov* (Princeton, 1952). Dashkova's uncle, Mikhail I. Vorontsov, had been Lomonosov's patron, and it is likely that Dashkova had met him when she was growing up in her uncle's house.

20. For Karzhavin, see S. R. Dolgova, *Tvorcheskii put' F. V. Karzhavina* (Leningrad, 1984).

21. Dashkova, *Memoirs*, p. 215.

22. See Dolgova, *Kniaginia E. R. Dashkova i sem'ia Malinovskikh*, and "E. R. Dashkova i sem'ia Malinovskikh" in *Ekaterina Romanovna Dashkova: issledovaniia i materialy*, ed. A. I. Vorontsov-Dashkov et al. (St. Petersburg, 1996), pp. 71–79.

23. Zuev was the author of several works on zoology and the first Russian textbook on natural sciences. He participated in Pallas's expeditions in 1768–74. In 1781–82, Zuev traveled to the lower Dniestr and Bug Rivers and described his travels in *Puteshetvennyia zapiski Vasilia Zueva ot S. Peterburga do Khersona v 1781 i 1782 godu* (St. Petersburg, 1787).

24. Archives of the Russian Academy of Sciences, PFA RAS, f. 3, op. 1, d. 556, l. 153 ob.

25. See Dolgova, *Kniaginia E. R. Dashkova i sem'ia Malinovskikh*, pp. 195–96.

26. Khrapovitskii, *Dnevnik A. V. Khrapovitskago*, pp. 277–78; C. F. P. Masson, *Mémoires secrets sur la Russie*, vol. 2 (Paris, 1800), pp. 142, 154.

27. Khrapovitskii, *Dnevnik A. V. Khrapovitskago*, p. 48.

28. See N.Yu. Bolotina, "Raznye zhizni sestyor Vorontsovykh: Ekaterina Dashkova i Anna Strogonova" in *E. R. Dashkova i eyo vremia: Issledovania i informatsiya* (Moscow, 2000), pp. 34–38.

29. *Russkii arkhiv,* no. 1 (1883), p. 224.

30. *Arkhiv kniazia Vorontsova,* ed. P. I. Bartenev (Moscow, 1872), vol. 5, p. 105.

31. Dashkova, *Memoirs*, p. 253.

"A Red Hot English Woman":

Princess Dashkova's Love Affair with Britain

ANTHONY CROSS

By the time Catherine II became empress of Russia in 1762, Anglomania had touched most of Europe and would gather momentum in the decades that followed. In 1758, toward the end of Empress Elizabeth's reign, an operatic version of Carlo Goldoni's play *La Ritornata di Londra*, in which the vogue for English fashions was ridiculed, was played in St. Petersburg to a restricted but appreciative audience, who was yet unable to enjoy the delights or resist the seductions of "blessed Albion" at first hand.[1] The opportunity for the Russian aristocracy and gentry to travel abroad without hindrance was one of the provisions of the major decree issued during the short reign of Catherine's husband, Peter III. Thus, during the 1770s and 1780s in particular, many Russians embarked on their version of the Grand Tour, which included visits to London and sometimes even the more remote regions of Britain. On two occasions Princess Ekaterina Romanovna Dashkova was among those Russians who made the journey to England.

Anglomania indicates an excessive attachment to all things English, mainly the more trivial and superficial, while Anglophilia suggests a more reasoned, discriminating attitude toward England with particular regard to its cultural heritage and social and political institutions. Dashkova would have undoubtedly placed herself in the latter category, the passion and outspokenness of her pronouncements about England throughout her adult life notwithstanding. As early as 1763, still only twenty years of age and already losing favor with Empress Catherine, whom she believed she had been largely responsible for putting on the throne, Dashkova was reported by the British ambassador as saying: "Why was I not born an Englishwoman? I adore the freedom and spirit of that nation."[2] Twenty years later, long after her visits to Britain, while living on her estate at Troitskoe (near Moscow), Dashkova was described by her young Irish companion Martha Wilmot as "a red hot English Woman."[3]

A member of the immensely powerful Vorontsov clan, Dashkova rivaled her two brothers, Aleksandr and Semion—who both served as Russian ambassadors in London—in their firsthand knowledge of and sympathy for England. Almost certainly, she excelled them in her ability to speak the language, although they all preferred to express themselves and their admiration for England in French.[4] In 1776 it was suggested: "Tho she does not speak English well, she understands it perfectly & converses without much embarrassment to herself in that Language."[5] Toward the end of her life a fellow of an Oxford college, who had met Dashkova in Russia, said she spoke "admirable English."[6]

A decisive moment in Dashkova's engagement with Britain came at Spa in Belgium, where she and her two young children, Anastasiia (b. 1760) and Pavel (b. 1763), had arrived in the summer of 1770 during their European travels. Diplomats apart, the princess had met few Britons during the first years of Catherine's reign, but at Spa she made important and long-lasting friendships, notably with two Irish ladies, Catherine Hamilton, the widowed daughter of the bishop of Tuam, and Elizabeth Morgan, daughter of the Irish attorney general. Crucial also was the meeting with John Hinchcliffe, who, before becoming bishop of Peterborough, had been headmaster of the Westminster School, a place he recommended as very suitable to educate young Pavel.

By October 1770, the Dashkovs were settled in London. Pavel did not enroll at Westminster but was left in the care of the Russian ambassador, Aleksei Semenovich Musin-Pushkin, and his wife, when his mother and sister departed on a tour that took them in the space of a fortnight to Bath via Portsmouth and Salisbury and on the return journey to Oxford and Windsor. It was a route that countless other tourists had followed, but as yet few Russians. Dashkova would be the first of them to publish an account of her travels: *Puteshestvie odnoi Rossiiskoi znatnoi Gospozhi po nekotorym Aglinskim provintsiiam* [The Journey of a Russian Noblewoman through Certain Provinces of England] is thus a landmark, albeit a modest one, in the history of Russian letters.[7]

To an admittedly small Russian readership, Dashkova used her travel journal to introduce her impressions of English cities and towns, particularly Bath and Oxford, where she stayed the longest. She created amusing pen pictures of such places as the famous Pump Room and adjacent baths in Bath, where "men and women are all mixed together and are dressed in yellow flannel garments with hats made from oilskin on their heads";[8] she also delivered incisive judgments on the cathedral at Salisbury, "an enormous Gothic building, remarkable only for its antiquity," and the great circle at Stonehenge.[9] Oxford clearly impressed her and she provided detailed descriptions of its colleges, libraries and museums (she also mentioned meeting a group of Russian students, who were studying there at the time of her visit). Above all, her journal is notable for its descriptions of great houses—such as Claremont, Longleat, and Blenheim—and their magnificent landscape gardens. It was at Painshill, the extraordinary garden of the Honorable Charles Hamilton, that she saw Joseph Lane's elaborate grotto. Her description of the place in its detail is without rival, including even English accounts. Visiting Wilton, the estate of Lord Pembroke, she could hardly imagine that some thirty years later, her niece, also a Catherine, the daughter of the Russian ambassador Semion Vorontsov, would marry into the British aristocracy and ride her Russian sleigh through those very grounds. Dashkova became as passionate a devotee of the English garden as Catherine the Great, stating: "their gardens alone are worthy of being described by a epic poet."[10] Later in life while improving her estate at Troitskoe in the English manner, she jokingly offered her services to her brother as "your English gardener."[11]

Dashkova's first sojourn in England gave concrete form to her previously abstract Anglophilia. In the introduction to her *Journey* she wrote: "England impressed me more than any other country. Their government, education, conduct, public and private life, machinery, buildings and gardens, all dependent on the form of the first of these, exceed the most strenuous efforts of other nations in similar undertakings."[12] Denis Diderot, the eminent *philosophe* whom she met in Paris after leaving England, had no doubts of her pro-English inclinations: "She has developed quite a taste for the English nation and I am a little afraid that her partiality for this antimonarchical nation prevents her from appreciating the advantages of mine."[13]

It was wholly appropriate that two years after her return to Russia, Dashkova would publish her travel journal in the Free Russian Society of Moscow University's almanac, which reflected her interests both in education and in England. Indeed, it was in the same collection in 1775 that her friend Mikhail Pleshcheev—who had served for a decade in the Russian embassy in London, initially under her brother Aleksandr— published his translation of Hamlet's monologue "To be or not to be," which he signed with the eloquent *nom de plume* "Angloman" (Anglomaniac). The following year, Pleshcheev published other items, including a poem entitled "Lines written by an Oxford Student on Looking at a Portrait of Locke."[14]

John Locke's ideas on education strongly influenced Dashkova's own,[15] but ultimately she would formally educate her son in Edinburgh. It was a decision dictated not only by reasons of economy (the Scottish universities were reputed to be far cheaper than Oxbridge), but also by the quality of Edinburgh University's teaching (Oxford, she noted, had "fallen far from what it was").[16] In 1776, Dashkova was writing to Edinburgh University's principal, William Robertson—known throughout Europe for his historical studies—outlining the course of education she envisaged for her son. In the end, they agreed on a compromise between what Dashkova wanted and what the university offered. Dashkova recalled the three years that she and her family spent in Edinburgh as "both the happiest and most peaceful that has ever fallen to my lot in this world;"[17] and as a mother who had "fulfilled the functions and duties of a tutor," she was inordinately proud of Pavel's success in achieving his master's degree in May 1779.[18]

During this second residence in Britain, she and her family also traveled widely, spending nearly a year in Ireland after leaving Scotland in June 1779 and before returning to London via Wales. Her brief impressions of Ireland are found in her memoirs, where she also alludes to a trip to the Scottish Highlands made during a university vacation. Unlike the description of her excursion in 1770, this second account, which took the form of a long letter to her friend Elizabeth Morgan, remained unpublished until very recently; however, it is no less unique in the Russian context (and not merely for the first attempt to describe the kilt), for it combines her admiration for the Scottish countryside, particularly Loch Lomond, with a clear awareness of the poverty and hardships of many of the inhabitants.[19]

In 1770 Dashkova had traveled to England incognito as Mme. Mikhalkova; in 1776 she returned to receive the open recognition her rank warranted, at least in her own eyes. During the earlier visit she had made many important contacts, but during her second longer sojourn from October 1776 to August 1780 she made the acquaintance of a vast number of Britons, from the royal family, through the echelons of the aristocracy and gentry, to a whole phalanx of the country's leading writers, scholars, and artists. Everywhere she went, she created a stir and prompted a flurry of letters and memoirs in which all aspects of her conduct, physical appearance, and dress were duly noted and frequently criticized.[20] Many of these comments came from women, including a whole clutch of Bluestockings, who were especially attentive to manifestations, sartorial and otherwise, of her masculinity. Some of the friendships she made, however, survived the years, as two very different instances demonstrate.

In the summer of 1780 on their return to London the Dashkovs visited Windsor and other estates in the vicinity of the capital. One such was Gregories at Beaconsfield in Buckinghamshire, home of the renowned parliamentarian, the Irishman Edmund Burke. Both Dashkova and her son Pavel—who had quoted Burke's *Philosophical Enquiry into the Origin of Our Ideas of the Sublime and the Beautiful* (1757) in his dissertation—[21] were highly impressed on meeting him and his wife. In a letter to Burke, written just before her departure from England, Dashkova extolled "Characters that do Honour to the Human Species."[22] A decade later, when Burke published *Reflections on the Revolution in France* (1790)—his famous and persuasive attack on the excesses of the new French "freedom" that was to find very sympathetic readers in Empress Catherine, Dashkova, and her brother Semion—she sent Burke her portrait, most probably J. C. Mayr's engraving from the portrait by Dmitrii Levitskii, depicting her wearing her star of the Order of St. Catherine and listing all her titles in the inscription.[23]

Another of Dashkova's long-term friendships came out of a chance encounter as the family traveled north from London to Edinburgh in November 1776. They stayed overnight at Easton Maudit, the country seat in Northamptonshire of the duke and duchess of Sussex, whom Dashkova had met during her first sojourn in Britain. It was at dinner that night that Dashkova met Edward Wilmot, yet another Irishman, and a relative of the Sussexes. It was his one-year-old daughter Martha, who would, some twenty-five years later, become the comfort and balm of Dashkova's old age and the catalyst to Dashkova writing her famous memoirs.[24]

Among the people she met, one group occupied a conspicuous place in her affections and her future activities: the Edinburgh professors, headed by William Robertson, to whom she wrote in 1786 to recall "the Epoque, when I had the happiness of your amiable and instructive Society, and regret the circumstances that prevent its repetition." She went on to emphasize that she was "so proud of the Esteem of my Scotch friends that I cannot give them leave to forget me."[25] In her memoirs, Dashkova painted in the rosiest of colors:

> the University professors, all of whom were generally esteemed for their intelligence, intellectual distinction, and moral qualities. Strangers alike to envy and to the pretentiousness of smaller minds, they lived together in brotherly amity, their mutual love and respect making of them a group of educated and intelligent people whom it was always an immense pleasure to see and whose conversation never failed to be instructive.[26]

Furthermore, she averred that "the immortal Robertson, Blair, Smith and Ferguson came twice a week to spend the day with me," which, if true, would have left them little time to produce the works that brought them fame.[27]

Here Dashkova named but a selection, however distinguished, of the professoriate that brought Edinburgh its unparalleled reputation during the last decades of the century under the dynamic leadership of Robertson. Her son had followed courses not only taught by Hugh Blair and Adam Ferguson but also John Bruce, Dugald Stewart, and John Robison (whom Dashkova had known in Russia, where he had been professor of mathematics at the Naval Cadet Corps at Cronstadt). Rhetoric, natural and moral philosophy, and mathematics, the subjects taught by these professors, were all outlined in the plan Dashkova submitted to Robertson regarding Pavel's education; he also studied chemistry with the renowned Joseph Black, to whom he later wrote to express his gratitude "for the many kind instructions and all the good offices I received from you."[28]

Evidence of Dashkova's continued esteem is not hard to find, especially after the empress appointed her director of the Imperial Academy of Sciences and Arts in St. Petersburg in January 1783. Among her first acts was the engineering of the elections of Robertson and Black as honorary foreign members. Moreover, Black was called upon to correspond with the Academy on chemical matters, a task he seems to have diligently fulfilled; for instance, in a letter of 1787 he informed Dashkova of "her friend" James Hutton's revolutionary theory of the origin of the earth's crust.[29]

In May 1783 Dashkova initiated the publication of a new journal entitled *Sobesednik liubitelei rossiiskogo slova* [The Interlocutor of Lovers of the Russian Word], to which she contributed a number of editorials and articles. Among the latter was "O smysle slova 'vospitanie'" [On the Meaning of the Word "Education"], in which Dashkova argues that the word in Russian has no specific meaning, but that true "education" should consist of physical, moral, and school (or classical) elements—the first two are essential for everyone, while the third is necessary for enriching the knowledge of "a certain rank." In the section of her article devoted to classical education Dashkova points directly to the beneficial example of English universities and of Edinburgh in particular, where:

> they examine much more strictly than elsewhere . . . in order to achieve the degree of Magister artium it is necessary to know the following sciences so well that at the public examination not only the questions of the professors but also of the attending public (because everyone has the right to put questions to the candidate) must be answered satisfactorily, and namely, logic, rhetoric, history, geography, higher mathematics, moral philosophy, jurisprudence, natural philosophy, which they began to teach in Edinburgh a few years ago based on mathematics.[30]

To the subject of logic she made a footnote, singling out John Bruce for the clarity of his teaching, free from obfuscating pedantry.

The year 1783 was in every sense a high point in expressions of Dashkova's positive attitudes to Britain. After a summer spent entertaining her friend Catherine Hamilton, who was visiting Russia for the first time, in October Dashkova became president of the new Russian Academy, which as her brainchild was dedicated to "the purifying and enrichment of the Russian language."[31] Among the initial academicians chosen for their knowledge of the Russian language were four men who also had a command of English and had also spent many years in Britain. Two, she had met at Oxford in 1770: Vasilii Nikitin and Prokhor Suvorov, who remained there until 1775, when they became the only Russians in the eighteenth century to earn Oxford degrees. The third was Semion Desnitskii, who had received his master's degree at the University of Glasgow before proceeding to his doctorate in law in 1767. The fourth member did not study at a British university but was the outstanding scholar and poet Vasilii Petrov, dubbed Catherine II's "pocket poet" for the involved eulogies he composed in her honor.[32] Dashkova's awareness of these scholars' long exposure to Britain and their linguistic abilities (Nikitin, Suvorov, and Desnitskii were all teaching English) was undoubtedly a significant factor in their election.

In her views on education, in her understanding of the workings of learned societies, in her enthusiasm for the English landscape garden, and in myriad other ways Dashkova was profoundly influenced by British example and achievement. Britain was thus no mere chapter in her life but a red thread in her activities and thinking. Thus, it was appropriate that she dictated her famous and controversial memoirs to a Briton, the Irishwoman Martha Wilmot, her companion in her last years in Russia.

Notes

1. P. N. Petrov, *Istoriia Sankt-Peterburga* (St. Petersburg, 1885), pp. 587–88.

2. *The Despatches and Correspondence of John, Second Earl of Buckinghamshire, Ambassador to the Court of Catherine II of Russia, 1762–1765*, ed. Adelaide D'Arcy Collyer, vol. 2 (London, 1902), p. 38 n. 2.

3. Martha and Catherine Wilmot, *The Russian Journals of Martha and Catherine Wilmot*, eds. Marchioness of Londonderry and H. M. Hyde (London, 1934), p. 50.

4. On the careers in England of Aleksandr, who served from February 1762 until April 1763, and of Semion, who was appointed in 1785 and lived in England until his death in 1832, see Anthony Cross, "*By the Banks of the Thames": Russians in Eighteenth-Century Britain* (Newtonville, Mass., 1980), pp. 16–18, 23–28.

5. National Library of Scotland, Edinburgh, MS 3942, f. 291. See also, Horace Walpole's opinion of her at the time of her first visit to London in 1770 (*Horace Walpole's Correspondence*, ed. W. S. Lewis, vol. 23 [New Haven, 1967], p. 249).

6. *Life of Reginald Heber, D.D. Lord Bishop of Calcutta. By His Widow. With Selections from His Correspondence, Unpublished Poems, and Private Papers; together with a Journal of His Tour in Norway, Sweden, Russia, Hungary and Germany, and a History of the Cossacks*, vol. 1 (London, 1830), p. 196.

7. *Opyt trudov Vol'nogo rossiiskogo sobraniia*, vol. 2 (Moscow, 1775), pp. 105–44.

8. *Opyt trudov*, vol. 2, p. 121.

9. *Opyt trudov*, vol. 2, p. 117.

10. *Opyt trudov*, vol. 2, p. 106.

11. *Arkhiv kniazia Vorontsova* (Moscow, 1882), vol. 21, p. 428.

12. *Opyt trudov*, vol. 2, p. 106.

13. Denis Diderot, "Sur la princesse Dashkoff," *Oeuvres complètes*, vol. 27 (Paris, 1876), pp. 248–49.

14. *Opyt trudov*, vol. 2, pp. 257–61; vol. 3, pp. 72–74. Pleshcheev, incidentally, left his signature in the admissions register of the Bodleian Library in Oxford in June 1762.

15. Dashkova, in a letter to Catherine Wilmot, published in *Drug prosveshcheniia* in 1806, refers to her early reading of Locke's *Some Thoughts on Education* in French translation (E. R. Dashkova, *O smysle slova "vospitanie": sochineniia, pis'ma, dokumenty*, ed. G. I. Smagina [St. Petersburg, 2001], p. 219).

16. *Opyt trudov*, vol. 2, p. 135.

17. E. R. Dashkova, *The Memoirs of Princess Dashkova*, trans. and ed. Kyril Fitzlyon (London, 1958), p. 147.

18. Dashkova, *Memoirs*, p. 149.

19. The French text, "Le petit Tour dans les Highlands" (with a Russian translation) forms an appendix to my article "Poezdki kniagini E.R. Dashkovoi v Velikobritaniiu (1770 I 1776–1780 gg.) i ee 'Nebol'shoe puteshestvie v Gornuiu Shotlandiiu' (1777)," *XVIII vek*, vol. 19 (1995), pp. 239–68.

20. See Anthony Cross, "Contemporary British Responses (1762–1810) to the Personality and Career of Princess Ekaterina Romanovna Dashkova," *Oxford Slavonic Papers*, n.s., vol. 27 (1994), pp. 41–61.

21. Paulus Dashkov, *Dissertatio philosophica inauguralis, de Tragoedia* (Edinburgh, 1779), p. 17. Burke is described as "scriptor pulcherrimus simul ac sagacissimus."

22. Sheffield City Archives, Wentworth Woodhouse Muniments, WWM Bk1/1401. See also the letter from Pavel Dashkov to Edmund Burke, September 19, 1780 (also in the Wentworth Woodhouse Muniments, Bk1/1361).

23. Archive of the Institute of History, Russian Academy of Sciences, St. Petersburg, Fond 36, Vorontsovy, op. I, d.1249, f. 82 (letter from Edmund Burke to S. R. Vorontsov, January 18, 1792).

24. Dashkova, *Memoirs*, p. 146.

25. National Library of Scotland, Ms. 3942, f. 265.

26. Dashkova, *Memoirs*, p. 147.

27. Among "the works that brought them fame" were William Robertson's *History of the Reign of the Emperor Charles V* (1769) and *History of America* (1777), both of which were translated into Russian during Catherine II's reign; Hugh Blair's *Lectures on Rhetoric and Belles Lettres* (1783); Adam Smith's *Theory of the Moral Sentiments* (1759) and *An Inquiry into the Nature and Causes of the Wealth of Nations* (1776); and Adam Ferguson's *Essays on the History of Civil Society* (1767) and *Institutes of Moral Philosophy* (1769).

28. Edinburgh University Library, Ms. Gen. 873/III, f. 90.

29. B. L. Modzalevskii, *Spisok chlenov Imp. Akademii Nauk* (St. Petersburg, 1908), p. 135; Edinburgh University Library, Ms. Gen. 873/III, ff. 36–8.

30. Dashkova, *O smysle slova "vospitanie,"* p. 126.

31. See the proposals she presented to Catherine II in August 1783 (Dashkova, *O smysle slova "vospitanie,"* pp. 287–88).

32. The careers of all four men, particularly their years in Britain, are investigated in detail in Cross, *"By the Banks of the Thames,"* pp. 98–116 (Nikitin and Suvorov), pp. 122–28 (Desnitskii), pp. 224–28 (Petrov).

Portraits of Princess Dashkova

ELENA IGOREVNA STOLBOVA

*A*lthough Princess Ekaterina Romanovna Dashkova lived in the eighteenth century, she was the very epitome of a modern-day woman—a manager, publisher, and businesswoman..Because of her bold personality and unusual accomplishments, especially in science and education, she fascinated many of her contemporaries, some of whom described her in their correspondence and other writings. Of her appearance Denis Diderot wrote the following:

> The princess . . . is no beauty. She is short, has a
> high, open forehead, chubby cheeks, deep-set
> eyes, which are neither large nor small, black
> eyebrows and hair, a slightly flattened nose,

a large mouth . . . straight neck . . . high bust, corpulent . . . lacking in grace and nobility . . . In December 1770, although only twenty-seven, she looked forty.[1]

The British ambassador to Russia, John Hobart, the second earl of Buckinghamshire, offered a contradictory description of Dashkova: "Princess D'Ashkow . . . has a remarkably good figure, and presents herself well . . . her countenance pleases and her manner is calculated to raise those sentiments she scarcely ever knew."[2] But English author Sir Horace Walpole later compared the princess to the queen of the Amazons, calling her a "tigress," a "Scythian heroine," and "an absolute Tartar."[3]

Several other descriptions of Dashkova note a masculine character. For example, Gavriil R. Derzhavin, the leading Russian poet of her time, chose Apollo as the god who matched her most closely. And the princess herself said she was a "cavalier."[4] Moreover, English author Elizabeth Carter wrote of Dashkova in 1770:

> She seems to be a most extraordinary genius. She rides in boots, and in all the other habilments of a man, and in all the manners and attitudes belonging to that dress . . . she likewise dances in a masculine habit, and I believe appears as often in it as in her proper dress.[5]

Given such differences of opinion about her looks, it is not surprising to find that Dashkova's appearance in her extant portraits is diverse and inconsistent, and that their iconography varies greatly. About thirty portraits were created during the princess's lifetime, and most of them cluster into one of three iconographic categories or types: bust-length court portraits; nearly full-length seated portraits showing her as an intellectual leader; and images commemorating her exile. These types or groups, to be described below, usually consist of several nearly identical images made by one or more artists. There are also, however, other oil paintings, miniatures, prints, and works on paper that do not fall into these three categories.

Chief among the atypical oil paintings of Dashkova are two youthful portraits, both displayed in a 2003 exhibition at the State Historical Museum in Moscow. The first had been considered a portrait of Prince Aleksandr A. Menshikov since 1898.[6] Menshikov was the son of Aleksandr D. Menshikov, famous friend and confidant of Peter the Great. In 1996, the present author claimed that the identity of the sitter was wrong, and that it was, in fact, the earliest known image of Dashkova.[7] The mistaken attribution had endured for nearly a century perhaps in part because the author of an 1898 museum catalogue had rejected as inauthentic an inscription on the back of the canvas that reads: "La Princesse Dachkoff, née Worontsoff, amie de l'Impératrice Catherine." On the one hand, the author had good reason to discount the inscription because the sitter dons the clothes and hairstyle of a young man. On the other hand, the sitter also wears the star of the Order of St. Catherine, which was established to honor women. The author must have known, however, that Menshikov was an exception in that he was the only man ever to receive the star, making it possible that the portrait was indeed a likeness of him.

Nonetheless, several clues support the assertion that the image is of Dashkova: the inscription; the star of the Order of St. Catherine, which she was awarded on June 28, 1762, just after the coup that brought Catherine II into power; and especially the facial features and clothing of the sitter. The yellow jacket or caftan closely matches the uniform of the Cuirassiers Regiment, commanded by her husband, Prince Mikhail I. Dashkov, in July of 1762. Dashkova herself held an honorary position in the same regiment.[8]

It was not unusual for Dashkova to wear men's clothing. According to her memoirs, she ordered a male suit from a tailor to use as a disguise in anticipation of a possible coup to overthrow Peter III. It was not ready in time, which aggravated her,[9] so she and Catherine borrowed old uniforms from two officers of the Preobrazhenskii Guards regiment so they could travel easily on horseback with troops who had switched their allegiance to the new Empress. After the coup, Dashkova paid calls on her relatives while still wearing the uniform, as she had no time to return home to change clothes. Dashkova wrote that she "looked like a boy of fifteen in my uniform," causing some confusion among the Empress's council. She still had the uniform on when Catherine granted her the Order of St. Catherine as a reward for her support.[10] Thus, this early portrait, though it shows Dashkova in the uniform of her husband's regiment, rather than the one she wore during the coup, was most likely painted soon after the event to commemorate Dashkova's role in bringing Catherine to the throne.

The second oil portrait at the State Historical Museum that does not fit into the established Dashkova types also shows her in her youth.[11] Painted in a naïve style, it betrays the hand of a self-taught artist, possibly a serf. For this reason, the image does not offer a realistic likeness. Yet despite the lack of realism, the facial features strongly suggest Dashkova. In addition, the presence of the star of the Order of St. Catherine on the sitter's bodice helps confirm the probable identity, especially because very few young women were awarded this honor. Aside from Dashkova, the only other woman at the time to receive the star was Princess Catherine von Holstein-Beck, a cousin of the empress, who was also decorated in 1762, at the age of twelve. Furthermore, although the provenance of the portrait is unknown, it is believed to have belonged to the Vorontsov family, and was probably one of the forty Vorontsov family portraits Dashkova bequeathed to her nephew, Mikhail S. Vorontsov.[12]

The other major likenesses of Dashkova can be grouped into three categories or types. Defining the first category is a bust-length court portrait, the most artistically significant likeness of Dashkova. Painted by Dmitrii G. Levitskii and signed and dated 1784, it is now in the collection of the Hillwood Museum and Gardens in Washington, D.C. This masterpiece, one of three versions by Levitskii, belonged to the princess until she presented it, along with other expensive gifts and souvenirs, to Martha Wilmot, an Irishwoman who lived with her near the end of her life and transcribed her memoirs. In 1808 Wilmot took the portrait with her when she returned to Ireland.[13] Dashkova is clearly idealized in this portrait. The artist makes his subject appear attractive and even regal, creating a regular, oval face, a soft half-smile, and large, expressive eyes. The sitter is not a seductive beauty here but a majestic stateswoman not unlike Catherine the Great herself.[14]

Another version of this image, probably also painted by Levitskii, is now in a private collection in Moscow. This painting is slightly larger than the Hillwood canvas but is almost identical in composition. The princess's clothes, the star of the Order of St. Catherine, and the miniature portrait of Catherine the Great awarded to ladies-in-waiting all coincide. Only the face is significantly different. The final known portrait belonging to Levitskii's iconographic type is in the Vorontsov Palace Museum in Alupka in the Crimea.[15] It once belonged to Dashkova's nephew, Mikhail S. Vorontsov. Everything coincides with the original Levitskii portrait, except for the color of the bodice.[16]

The second major Dashkova portrait type shows a nearly full-length, seated figure of the princess in her role as director of two imperial academies: the Imperial Academy of Sciences and Arts in St. Petersburg and the Imperial Academy of the Russian Language. Included are a globe in the lower right and several volumes of the dictionary she published for the latter institution. Such accoutrements of knowledge were common in contemporaneous portraits of male scholars, but very rare in images of women.

The authorship of this portrait type is unknown, yet it is one of the most copied of all the oil portraits of Dashkova. There are four versions from the late eighteenth or early nineteenth centuries and two copies made very recently.[17] One of the period portraits, currently on loan to the State Hermitage Museum, belongs to the Pushkin House (Institute of Russian Literature) in St. Petersburg;[18] another belongs to the State Historical Museum in Moscow;[19] a third hangs in the presidium of the Russian Academy of Sciences in Moscow; and the fourth, one of the portraits of Dashkova taken to England by Martha Wilmot, hangs in Longleat House in Wiltshire, England.[20] The chronology of the four portraits, like the authorship, is unknown, but the original one was probably painted in the first half of the 1790s, or slightly earlier, perhaps to commemorate the publication of the dictionary for the Academy of the Russian Language.

The two modern versions of this iconographic type have been painted very recently. Elena Belova-Romanova produced a slightly different image for the Ekaterina Dashkova Humanitarian Institute in Moscow, and the American Philosophical Society in Philadelphia owns a copy of the State Historical Museum portrait painted by Alexei Nesterovich Maximow in 2003.

Images of Dashkova in exile constitute the third and final portrait type. After Catherine the Great died in November 1796, her son Paul I exiled the princess to the small village of Korotovo in the forests of Novgorod Province. She lived in a peasant hut, with only her books and a few close friends and relatives for companionship. This harsh exile lasted only a short time, and Dashkova was soon allowed to return to her estate of Troitskoe, near Moscow, though she remained excluded from court life. A few years later, in 1801, she commissioned Italian artist Salvator Tonci to execute this portrait in memory of this time.[21] There are several versions of this type. One, from the collection of Count A. L. Santi, is now in the Hermitage.[22] Another belongs to the State Historical Museum in Moscow.[23] A similar composition is encountered in the portrait from Martha Wilmot's album of watercolors known as the "Green Book."[24] In an engraving of this portrait, published in the first English and Russian editions of Dashkova's memoirs (1840 and 1856, respectively), a dog is added, lying at the princess's feet.

A bust-length version of this portrait type also exists. It is in a private collection in the United States. Other works that derive from the exile portrait include a watercolor copy, possibly based on an engraving, in the Vorontsov Palace Museum in Alupka; a large grisaille bust of the princess at the State Historical Museum in Moscow, which copies the head of the Hermitage exile portrait but changes other aspects of it; and finally, a copy by Viktor Vasnetsov that was made in the late nineteenth or early twentieth centuries for the gallery of distinguished eighteenth-century figures in the Rumiantsev Museum in Moscow.

In addition to the two youthful oil portraits discussed above that do not belong to any of the three major iconographic types, there are a number of other atypical depictions, including prints, paintings, miniatures, and works on paper. One of the best-known prints is a bust-length stipple engraving by Gavriil I. Skorodumov from 1777. This likeness, based on the artist's own drawing, was made during her second trip to Britain with her children (1776–80). The portrait circulated widely in print form and was later copied by other artists. Prints made from the plate in London bear the date of October 22, 1777, a coat of arms, and an inscription that reads, "Her Highness the Princess of Dashkaw," along with the following lines, "Polite as all her Life in Courts had been / Yet Good as She the World had never seen." Skorodumov also made an unfinished family-portrait engraving that shows Dashkova with her son and daughter.

Another depiction of Dashkova made during her British sojourn deserves mention. For several years, Dashkova lived in Edinburgh where her son Pavel was studying for his diploma at the University. In addition to being presented to King George III in London, the princess traveled to the Scottish Highlands and to Ireland, where she most enjoyed her stay in Dublin in 1779 and 1780.[25] There her presence was recorded for posterity in a large painting by Francis Wheatley entitled *The Dublin Volunteers on College Green, 4th November 1779*, now in the collection of the National Gallery of Ireland. The painting depicts soldiers on foot and horseback on a square in Dublin on a bright, sunny day. All the faces are clearly portraits. Among the ladies looking out of the windows of a neighboring house is one wearing the red ribbon of an order and holding a parasol. There is good reason to believe that this is an image of Dashkova, especially since a key to the painting published later identifies her along with other prominent figures.[26] Dashkova did not mention this particular episode in her memoirs, but she would undoubtedly have enjoyed such a lively spectacle. As she wrote: "Dublin Society was then distinguished by its elegance, its wit, and its manners, and enlivened by that frankness which comes naturally to the Irish."[27]

There are also at least three miniature likenesses of Dashkova that are also unique. Two belong to the State Historical Museum in Moscow. One, painted in an academic style, is by an unknown artist, and the other,[28] shown in the museum's 2003 exhibition (though not included in the catalogue), contains one unusual detail—the ribbon of the Order of St. Catherine is worn over the left shoulder. A third miniature, recently acquired by the Hermitage, was painted by British artist Ozias Humphrey in 1770. It is set in an elegant gold-mounted lacquer *carnet,* or notebook, made in Paris by Jean-Jacques Prevost.

Other images that are unique include an engraving not mentioned in Dmitrii A. Rovinskii's dictionary of Russian engravers. Owned by the Pushkin House in St. Petersburg, this picture is so unlike all other images of the princess that only the inscription (if it is accurate) allows us to identify her. Another engraving, which recalls the style of popular Russian *lubok* prints, depicts Dashkova in all her regalia and includes lines of poetry in French. There are two other known images, both of which were displayed in an exhibition of Russian art in London in 1935. One of them, an oil painting on parchment, is signed and dated *Maurice 1771.*[29] Brought to England by Martha Wilmot, it belonged to a Mrs. E. M. Walker until 1935. Unfortunately, it was not reproduced in the catalogue of the exhibition and its current whereabouts are unknown. Finally, one other portrait was executed in pastel by the English artist Daniel Gardner, when Dashkova visited Britain for the second time.[30]

Despite all the portraits thus far discussed, the only image mentioned by Dashkova in her memoirs is a bronze bust sculpted in Paris by Jean-Antoine Houdon. The princess wrote of this bust:

> I could not help remarking on seeing it that French artists had too
> much taste to make a likeness of me, for he would not let me be
> as God had shaped me, but had made a French duchess out of me,
> wearing a dress with a low-cut neck, instead of the simple, unaffected
> woman I really was.[31]

The bust was shown in the Paris Salon of 1783, listed as a portrait of "Madame la Princesse Dashkow, Directrice de l'Académie des Sciences de Saint-Petersbourg."[32] There is no information on its current location.

The number of portraits that were made of Dashkova is extraordinary, given that she was not an empress. They offer material proof of her status as an accomplished woman. Indeed, few portraits of women at the time show them in positions of power, especially in leadership roles such as the one Dashkova held at the Academy of Sciences. Dashkova's distinction in eighteenth-century Russia, though sometimes clouded during her lifetime by the vagaries of personal relationships and court intrigue, was recognized by the mid-nineteenth century. An 1870 exhibition of Russian portraits that included nearly a thousand portraits of leading historical figures is a case in point. Rather than being placed in the section titled "Women Adorning Society," Dashkova's likeness appeared in "Leading Historical Personalities" alongside military commanders, diplomats, writers, painters, and actors. She was the only woman.[33] Today, as in the nineteenth century, this talented, ambitious, and accomplished woman of many guises continues to pique our inteest.

Notes

1. Denis Diderot, "Sur la princesse Dashkoff," *Oeuvres complètes* (Paris, 1876), vol. 17, pp. 487–90. A few changes and additions were made to this essay by the editor as a way of providing background and context for an English-speaking audience. The essay was translated by Kenneth MacInnes.

2. John Hobart, *The Despatches and Correspondence of John, Second Earl of Buckinghamshire, Ambassador to the Court of Catherine II of Russia, 1762–1765,* ed. Adelaid D'Arcy Collyer (London, 1900), vol. 1, p. 99; quoted in Anthony Cross, "Contemporary British Responses (1762–1810) to the Personality and Career of Princess Ekaterina Romanovna Dashkova," in *The Semiotics of Russian Cultural History,* eds. Alexander D. Nakhimovsky and Alice Stone Nakhimovsky (Ithaca, N.Y., 1985), p. 44.

3. Horace Walpole, *The Yale Edition of Horace Walpole's Correspondence,* ed. W. S. Lewis et al. (New Haven, 1937–), vol. 33, p. 173; vol. 29, pp. 59, 60; vol. 23, p. 248.

4. E. R. Dashkova, *Zapiski knyagini Dashkovoi* (St. Petersburg, 1907), p. 318.

5. Elizabeth Carter, *Letters from Mrs. Elizabeth Carter to Mrs. Montague, Between the Years 1755 and 1800, Chiefly on Literary and Moral Subjects* (London, 1817), pp. 88–89.

6. *Moskovskii glavnyi arkhiv Ministerstva inostrannykh del: Portrety i kartiny, khranyaschiesya v nem,* 1st ed. (Moscow, 1898), p. 64.

7. E. I. Stolbova, "Dva portreta: novoie v ikonografii E. R. Dashkovoi," in *Ekaterina Romanovna Dashkova: issledovaniia i materialy,* eds. A. Woronzoff-Dashkoff and M. M. Safonov (St. Petersburg, 1996), pp. 176–80.

8. Stolbova, "Dva portreta," p. 177.

9. "To my great disappointment I learned from my maid that my tailor had not yet brought my suit of man's clothes" (E. R. Dashkova, *The Memoirs of Princess Dashkova,* trans. and ed. Kyril Fitzlyon [Durham, 1995], p. 76).

10. Dashkova, *Memoirs,* pp. 78–79, 83–86.

11. No. 11-5186; acquired prior to 1951 and reproduced in the exhibition catalogue *Sei put' tebe prinadlezhit: Kniagine Ekaterine Romanovne Dashkovoi posviaschaetsa* (Moscow, 2003), pp. 86–87, cat. 281.

12. The portraits were later inherited by his son, who died childless in 1882. His widow, Princess Maria Trubetskaya, left Russia to live in Italy. After her death, her property in Moscow was sold at auction.

13. It was shown in a 1935 exhibition of Russian art from private collections in Western Europe. The exhibition catalogue stated that it was lent by persons empowered to act for the late Mrs. E. M. Walker from Oxford; see *Catalogue of the Exhibition of Russian Art, 1 Belgrave Square, London S.W. 1, 4 June to 13th July, 1935* (London, 1935), no. 127. The portrait was later reproduced on page 215 of Tamara Talbot Rice's *Concise History of Russian Art* (New York, 1963), which listed it as part of the collection of Mrs. Herbert A. May in Washington, D.C. (p. 281). By 1971, according to an article in *The Connoisseur,* it belonged to Marjorie Merriweather Post, and it is now part of the collection of Hillwood Museum and Gardens, also in Washington, D.C. See Alan Bird, "Eighteenth-Century Russian Painters in Western Collections," *The*

Connoisseur, vol. 178 (October 1971), p. 81; and Anne Odom and Liana Paredes Arend, *A Taste for Splendor: Russian Imperial and European Treasures from the Hillwood Museum* (Alexandria, Va., 1998), pp. 206–7, cat. 101. Post was an art collector who lived in Moscow from 1936 to 1938 as the wife of the American ambassador, Joseph E. Davies. Her unique collection of Russian art later formed the basis of the Hillwood Museum and Gardens in Washington, D.C.

14. Johann Conrad Mayr's engraving of Dashkova may have been made from this portrait.

15. No. 157 (oil on canvas, 25.3 x 20.2 inches [64.5 x 51.5 cm]).

16. This work was first published in 1984 in Vasyl′ V. Ruban's book on Ukrainian portraits of the first half of the nineteenth century; see V. V. Ruban, *Ukrainsk'ii portretnii zhivopis pershoi polovini XIX stolittia* (Kiev, 1984), p. 188. The copy shown at the *Historical Exhibition of Portraits* in St. Petersburg in 1870 (see note 33 below) may have been based on this work, rather than the Levitskii portrait. Another portrait of Dashkova by Levitskii is mentioned in the catalogue of an exhibition of historical works held at the Stroganov Art School in Moscow in 1901; see *Vystavka khudozhestvennyh proizvedeni stariny v Stroganovskom uchilische* (Moscow, 1901), cat. 601.

17. E. P. Renne, "Portret E. R. Dashkovoi v chastnom sobranii v Anglii," in *Vorontsovy-dva veka v istorii Rossii,* edition 7 (St. Petersburg, 2002), pp. 79–82.

18. No. 48952/6 (oil on canvas, 45 ¼ x 34 inches [115 x 86.5 cm]).

19. No. II-5469/52044/1919 (oil on canvas, 47 ½ x 36 ¾ inches [121 x 93.5 cm]).

20. An engraving based on this portrait, which is the smallest of the four, was published in volume 1 of the first edition of Dashkova's memoirs (London, 1840). The version belonging to the State Historical Museum has a signature—"Antropov 1784"—but it has been proven false. The English work has been attributed to Joseph Grassi or Marcelli Bacciarelli (who worked in Poland), though neither seems likely.

21. *Katalog vystavki portretov russkikh dostoprimechatel′kh liudei na postoiannoi vystavke . . . Obschestva liubitelei khudozhestv* (Moscow, 1868), p. 43, no. 82.

22. No. RZh-1884 (oil on canvas, 18 x 14 inches [46 x 36 cm]).

23. No. II-1606/70899 (oil on canvas, 23 1/8 x 18 ½ inches [59 x 47 cm]).

24. Diana Scarisbrick, "Companion to a Russian Princess: Martha Wilmot's Green Book," *Country Life,* vol. 169 (8 January 1981), p. 76.

25. A. G. Cross, "Poezdki kniagini E. R. Dashkovoi v Velikobritaniiu (1770 i 1776–1780 gg.) i ee 'Nebol'shoe puteshestvie v Gornuiu Shotlandiiu' (1777)," *XVIII vek,* vol. 19 (1995), p. 224.

26. This work was first brought to my attention by Elizabeth Renne of the State Hermitage Museum. See Anne Crookshank and the Knight of Glin, *The Painters of Ireland, c. 1660–1920* (London, 1978), pp. 152–53; and Mary Webster, *Francis Wheatley* (London, 1970), pp. 34, 126–27.

27. Dashkova, *Memoirs,* p. 189.

28. No. IV-485 (watercolor and gouache on bone, 1 ¾ x 1 ½ inches [4.5 x 3.8 cm]).

29. *Catalogue of the Exhibition of Russian Art,* cat. 814.

30. *Catalogue of the Exhibition of Russian Art,* cat. 669.

31. Dashkova, *Memoirs,* p. 158.

32. *Explication des peintures, sculptures et gravures . . .* (Paris, 1783), p. 49; reprinted in *Paris Salons de 1775, 1777, 1779, 1781, 1783* (New York, 1977).

33. *Katalog vystavki russkikh portretov izvestnykh lits XVI–XVIII vekov, ustroennoi Obschestvom pooshchreniia khudozhnikov* (St. Petersburg, 1870); and *Vystavka Obschestva pooshchreniia khudozhnikov: Istoricheskii al′bom portretov izvestnykh lits XVI–XVIII vekov, fotografirovannyi i izdannyi khudozhnikom A. M. Lushevym* (St. Petersburg, 1870), p. 36.

Author Biographies

Michael D. Gordin is an Assistant Professor in the History Department at Princeton University, specializing in the history of the modern physical sciences and the history of Imperial Russia. He has published widely on the history of Russian science from the early eighteenth century until the end of the Soviet period. His cultural biography of Russian chemist D. I. Mendeleev, noted for his 1869 formulation of the periodic system of chemical elements, appeared from Basic Books in 2004 as *A Well-Ordered Thing: Dmitrii Mendeleev and the Shadow of the Periodic Table*. He is currently working on a book on the history of the atomic bombing of Japan during World War II.

Michelle Lamarche Marrese is the author of *A Woman's Kingdom: Noblewomen and the Control of Property in Russia, 1700–1861*. She has taught at the University of Delaware, Northwestern University, and the University of Toronto, and is currently working on two projects: a study of the impact of female rule on Russian society in the eighteenth century and a biography of Dashkova.

Marcus Levitt is Associate Professor and Chair of the Department of Slavic Languages and Literatures at the University of Southern California. He is author of *Russian Literary Politics and the Pushkin Celebration of 1880*, editor of *Early Modern Russian Writers, Late Seventeenth and Eighteenth Centuries*, and co-editor of *Eros and Pornography in Russian Culture* together with A. Toporkov. He has written many articles on eighteenth- and nineteenth-century Russian literature. His current research investigates the status of the visual in eighteenth-century Russian culture.

Karen Duval is Associate Editor at the Papers of Benjamin Franklin, where she has worked since 1990. She has principal responsibility for annotating Franklin's French correspondence, in particular his exchange of letters with fellow freemasons, scientific colleagues, friends, and neighbors in France, as well as a range of importuning favor-seekers in all countries. She has a B.A. and M.A. in French literature from George Washington University and has done doctoral work in sixteenth-century French literature at New York University.

Alexander Woronzoff-Dashkoff is a Professor of Russian at Smith College. For many years he also worked in the Russian School at Middlebury College, the last nine years as Director of the school. His scholarship has been devoted to the life and works of Dashkova, of whom he is a descendent. Recently, he compiled and annotated the French edition of her autobiography, *Mon histoire*, and is currently writing her biography.

Svetlana Romanovna Dolgova is the Head of the Publications and Usage Department of the Russian State Archive of Ancient Acts in Moscow; her area of expertise is eighteenth-century Russia. She is the author of *Kniaginia E.R. Dashkova and sem'ia Malinovskikh* ("*Princess E.R. Dashkova and the Malinowski Family*") and *Korotkie rasskazy o Moskve* ("*Short Stories about Moscow*") and of numerous journal articles on the eighteenth century. She is one of the founders of the Worontsov society and is the Vice-Chair of the Book Section of the Moscow Scientific Society.

Anthony Cross is Professor Emeritus of Slavonic Studies at the University of Cambridge and a Fellow of the British Academy. He has written widely on eighteenth-century Russian culture and on Anglo-Russian cultural relations from the sixteenth century to the present. His books include "*By the Banks of the Thames*": *Russians in Eighteenth-Century Russia* and "*By the Banks of the Neva*": *Chapters from the Lives and Careers of the British in Eighteenth-Century Russia*. For the second of these he was awarded the 1998 Antsiferov Prize for the Best Foreign Work on St. Petersburg.

Elena Igorevna Stolbova is a Senior Research Specialist and Curator in the Department of Painting at the Russian State Museum. She has published articles and co-authored catalogues on painting of the eighteenth and early nineteenth centuries. She is a member of the Worontsoff society. Her research on the iconography of Dashkova's portraits began in 1987. She contributed to a five-series television documentary on the life of Dashkova, which is expected to air in 2006.

Checklist of the Exhibition

*O*bjects in this checklist are listed alphabetically by artist, author, or maker, with anonymous works at the beginning. Firms and institutions are included in the alphabetical listing. For published books and pamphlets, titles have been truncated, capitalization of titles has been regularized, place names have been anglicized, and publishers' names have been standardized. For published items (including prints), full publication information is given when available. Transliteration of Russian names and titles follows the Library of Congress system, with some adaptations.

Lenders to the Exhibition (with abbreviations used in checklist)

All objects exhibited belong to the American Philosophical Society (APS) unless otherwise noted.

APS at ANSP	American Philosophical Society, on deposit at the Academy of Natural Sciences, Philadelphia
Bakken	The Bakken, Minneapolis
BPL	The Boston Public Library, Rare Books Department, Courtesy of the Trustees
FI	The Franklin Institute Science Museum, Philadelphia
FLP	Free Library of Philadelphia, Map Collection
GIM	The State Historical Museum, Moscow
Hermitage	The State Hermitage Museum, St. Petersburg
Hillwood	Hillwood Museum and Gardens, Washington, D.C.
HSP	The Historical Society of Pennsylvania, Philadelphia
Kunstkamera	Peter the Great Museum of Anthropology and Ethnography (Kunstkamera) of the Russian Academy of Sciences, St. Petersburg
LC	The Library of Congress, Washington, D.C.
LCP	The Library Company of Philadelphia
Lilly	The Lilly Library, Indiana University, Bloomington
LWF	"Lest We Forget" – Museum of Slavery, Private Collection of J. Justin & Gwen Ragsdale
NYPL	The New York Public Library, Slavic and Baltic Division
PC	Private Collection
PHS	The Pennsylvania Horticultural Society, Philadelphia
PMA	Philadelphia Museum of Art
RAS	St. Petersburg Branch of the Archives of the Russian Academy of Sciences
RIA	Royal Irish Academy, Dublin
Rosenbach	The Rosenbach Museum & Library, Philadelphia
SCHS	Salem County Historical Society, Salem, N.J.
UAF	University of Alaska Fairbanks, Elmer E. Rasmuson Library, Rare Maps
UMich	University of Michigan, University Library, Ann Arbor
Walters	The Walters Art Museum, Baltimore

Sources of Facsimiles

BL	The British Library, London
NGI	National Gallery of Ireland, Dublin

NLS   The National Library of Scotland, Edinburgh

Yale   Yale University, The Beinecke Rare Book & Manuscript Library

Artist or Maker Unknown

Paintings

Portrait of Ekaterina R. Dashkova as a Young Woman, ca. 1762. Oil on canvas. GIM

Portrait of Ekaterina R. Dashkova in Military Uniform, ca. 1762. Oil on canvas. GIM

After Salvator Tonci. *Portrait of Ekaterina R. Dashkova in Exile,* late eighteenth–early nineteenth century. Oil on canvas. PC

After Mason Chamberlin. *Miniature Portrait of Benjamin Franklin,* early nineteenth century (after 1762 oil painting). Opaque watercolor on ivory.

Prints and Maps

After Benjamin West. *Benjamin Franklin Drawing Electricity from the Sky,* n.d. (after ca. 1805 oil painting). Color reproduction.

Benjamin Franklin, ministre plenipotentiaire à la cour de France, ca. 1780. Engraving.

Duc de Chaulnes's Improvement of Dr. Franklin's Electrical Kite, before 1788. Etching.

Allarme générale des habitants de Gonesse occasionée par la chute du ballon aréostatique de Mr. de Mongolfier. Augsburg, 1783. Colored etching. LC

Descente de la Machine Aerostatique des Srs. Charles et Robert, 1783. Colored etching. LC

Exp[é]rience a[ë]rostatique faite Versailles le 19 sept. 1783, 1783. Colored etching. LC

Experience de la machine areostatique, 1783. Colored etching. LC

Experience du globe aerostatique de MM. Charles et Robert au Jardin des Thuileries le 1er decembre 1783. Paris: Esnauts et Rapilly, 1783. Colored etching. LC

Decorative Arts

After Jean-Baptiste Nini. *Medallion Portrait of Benjamin Franklin,* late eighteenth century (after 1779 original). Painted terracotta.

Gold Box set with Turquoise Enamel Medallion Bearing the Cipher of Catherine II in Diamonds. Medallion, late eighteenth century; gold box, eighteenth or nineteenth century. Hillwood

Grand Cross (Znak or Krest) of the Order of St. Catherine, late eighteenth century. Gold, silver, enamel, and diamonds. Hillwood

Round Box with Catherine II as Minerva, 1781–82. Gold and verre eglomisé. Hillwood

Travelling Writing Case, before 1803. Red leather, embroidered with metal thread. RIA

Uniform of the Preobrazhenskii Regiment, 1762. Wool and gold braid. GIM

Scientific Instruments
Leyden Jar, second half of eighteenth century. Glass and metal. Bakken

Static (Frictional) Electrical Machine, late eighteenth century. Wood, brass, and glass. Kunstkamera

Burning Glass (Lens for Concentrating Sunlight), late eighteenth century. Glass lenses in wood and kidskin mount. Kunstkamera

Round Glass for Viewing Eclipse, eighteenth century. Green glass disc. Kunstkamera

Other Objects
Membership Token (Darik) for the Imperial Academy of the Russian Language, 1780s. Bronze. PC

Membership Token (Darik) for the Imperial Academy of the Russian Language, 1784. Bronze. RAS

Miniature Chess Set, eighteenth century. Ivory pieces with silver container and folding leather board.

Slave Ankle Irons, late eighteenth century. Iron. LWF

Slave Ankle Irons with Chain, late eighteenth century. Iron. LWF

Slave Branding Iron with Initials TC, late eighteenth–early nineteenth century. Iron. LWF

George Adams
An Essay on Electricity: Explaining the Theory and Practice of That Useful Science. London: Logographic Press, 1785.

American Philosophical Society
Magellanic Premium Medal, 1887. Gold.

Membership Diploma for Ekaterina R. Dashkova, 1791 (Russian copy of lost 1789 original). Manuscript. RAS

Johann Friedrich Anthing
Silhouettes of Russian Academicians Placing Leonhard Euler's Bust on a Pedestal, 1784. Cut paper and ink with gold leaf. RAS

Silhouettes of Russian Academicians Reading Under the Tree, 1784. Cut paper and ink with gold leaf. RAS

John Augustus Atkinson

Panoramic View of St. Petersburg: View from Vasil'evsky Island to the Palace Embankment, 1805–7. Colored etching and aquatint. LC

Panoramic View of St. Petersburg: View of the Twelve Colleges Building, 1805–7. Colored etching and aquatint. LC

Béry, Faiseur de Caractères

Group of Brass Stencils from Benjamin Franklin's Stencil Set, ca. 1781. Brass.

Stencil Specimen Sheet, ca. 1781. Stenciled sheet.

William Russell Birch and Thomas Birch

The City of Philadelphia: State House Garden; engraved 1798, published 1800 (1st ed.). Colored engraving and etching. LCP

The City of Philadelphia: Pennsylvania Hospital; engraved 1799, published 1800 (1st ed.). Colored engraving and etching. LCP

The City of Philadelphia: New Market; engraved 1799, published 1800 (1st ed.). Colored engraving and etching. LCP

The City of Philadelphia: Back of the State House; engraved 1800, published 1827–28 (4th ed.). Colored engraving and etching. LCP

Bowles and Carver

Bowles's New Pocket Map of the Discoveries Made by the Russians on the Northwest Coast of America. London: Bowles and Carver, 1770 (?). Colored engraving. UAF

Martha (Wilmot) Bradford

Two Journals, 1803 and 1805. RIA

Notebook of Russian Peasant Songs, n.d. Notebook covered in yellow-green silk. RIA

Document Envelope, n.d. Green silk with cord, tassels, and appliqués. RIA

Samuel Breck

Historical Sketch of the Continental Bills of Credit, 1840. Bound manuscript volume.

William L. Breton

Stokes's Old London Coffee House, 1830. Watercolor, graphite, and ink on paper. HSP

Joseph [?] Carmine

View of the Imperial Academy of Sciences and the Kunstkamera, late eighteenth century. Colored engraving. RAS

Catherine II, Empress of Russia

Manuscript Volume of Decrees (Nakaz), 1768. RAS

Daniel Chodowiecki

Franklin Before the French King. Etching in M. C. (Mathias Christian) Sprengel, *Allgemeines historisches taschenbuch.* Berlin: Haude und Spener, 1784.

Facsimiles of *Twelve Episodes from the American Revolution.* Etchings in M. C. (Mathias Christian) Sprengel, *Allgemeines historisches taschenbuch.* Berlin: Haude und Spener, 1784.

Ekaterina Romanovna Dashkova
Manuscripts

Facsimile of *Letter to William Robinson,* London, October 9, 1776. NLS

Letter to Benjamin Franklin, Paris, January 24, 1781.

Note to Benjamin Franklin, Paris, January 30, 1781.

Letter to Jonathan Williams, St. Petersburg, February 10/21, 1793. Lilly

Letter to the Village Head of Korotova, May 19, 1807. GIM

Epître Dédicatoire à Mademoiselle M. de Wilmot, 1805. Manuscript. RIA

Facsimile of *First Page of Manuscript Memoirs,* 1805. BL

Facsimile of *Copy of Letter to Mrs. Hamilton,* n.d. BL

Printed Works

Facsimile of *Recueil des Airs (Musical Compositions).* Edinburgh: Jaques Johnson, [1777?]. BL

Toisiokov (Mr. This-and-That or *Mr. Neither-Nor), Comedy in Five Acts.* Published in *Russian Theater (Rossiiskii featr),* part 19, vol. 9. St. Petersburg, 1788. RAS

Memoirs of the Princess Daschkaw, Lady of Honour to Catherine II, Empress of all the Russias. 2 vols. London: Henry Colburn, 1840.

Mémoires de la Princess Daschkoff, dame d'honneur de Catherine II, Impératrice de toutes les Russies. Paris: A. Franck, 1859. UMich

David

After Lejeune. Facsimile of *Franklin s'oppose aux taxes en 1766*, n.d. Engraving.

Nicolas Delaunay

After Ch. de Lorimier. *Premier voyage aërien en présence de Mr. le Dauphin. Expérience faite dans le Jardin . . . Vue de la Terrasse de M. Franklin à Passi*, 1783. Colored etching. LC

Denis

After Desrais. *Globe aërostatique. Dédié à Monsieur Charles. Cette machine est représenté ici s'élevant pour la seconde fois au milieu de la Praïrie de Nesle, ou il venoit de descendre, accompagné de Mr. Robert*, 1783. Colored etching. LC

Benjamin Eastburn

A Plan of the City of Philadelphia, the Capital of Pennsylvania. London: Andrew Dury, 1776. Engraving.

Sir William Elford

Plan of an African Ship's Lower Deck, with Negroes . . . Published by Order of the Pennsylvania Society for Promoting the Abolition of Slavery, 1789. Broadside.

Flugel

After I. K. Fioderov. Facsimile of *Catherine II and Ekaterina R. Dashkova Visiting Mikhail V. Lomonosov*. Wood engraving from *Neva* magazine, after oil painting, 1859. Kunstkamera

Benjamin Franklin
Manuscripts

Design for Currency, on verso of *Resolution for the Maintenance of Commerce, Proposed Resolution of the Continental Congress*, [July 1776?].

Designs for an Emblem of the Thirteen Original States, n.d. Ink and pencil on paper.

Fragment of *Letter to Georgiana Shipley*, Paris [?], 1781. LC

Letter to Ekaterina R. Dashkova, Philadelphia, May 7, 1788. LC

Letter to Mary (Polly) Stevenson, Paris, September 14, 1767.

Facsimile of *Letter to Ezra Stiles*, London, July 6, 1765. Yale

Passport for George F. Norton of Virginia, May 18, 1779.

Passport for the Ship "Harmony" of Bordeaux, August 29, 1777.

Printed Works

"Richard Saunders" [Benjamin Franklin]. *Poor Richard, 1735: An Almanack for the Year of Christ, 1735.* Philadelphia: Franklin, [1734]. Unbound pamphlet.

Plain Truth, or Serious Considerations on the Present State of the City of Philadelphia, and Province of Pennsylvania. [Philadelphia]: Franklin, 1747.

Experiments and Observations on Electricity, Made at Philadelphia in America by Mr. Benjamin Franklin. London: E. Cave, 1751.

Experiments and Observations on Electricity, Made at Philadelphia in America by Mr. Benjamin Franklin. London: David Henry, 1769.

Facsimile of *Illustration of Stoves,* from *Experiments and Observations on Electricity, Made at Philadelphia in America by Mr. Benjamin Franklin.* London: David Henry, 1769.

Passport for British Ships Returning American Prisoners of War to the United States, 1782.

With Timothy Folger. *Remarques sur la navigation de Terre-Neuve à New York afin d'éviter les Courrants et les bas-fonds au sud de Nantuckett et du Banc de George.* Paris: La Rouge, ca. 1785.

"The Morals of Chess." From *The Columbian Magazine,* vol. 1 (December 1786).

An Address to the Public. Philadelphia: Francis Bailey, [1789?]. Broadside.

Mémoires de la vie privée de Benjamin Franklin, écrits par lui-même, et addressés à son fils. Paris: Buisson, 1791.

The Private Life of Benjamin Franklin. London: J. Parsons, 1793.

The Life of Benjamin Franklin, Written by Himself. New York: S. King; printed by Clayton & Van Norden, 1824.

Memoirs of Benjamin Franklin, Written by Himself and Continued by His Grandson and Others. Philadelphia: M'Carty and Davis, 1834.

The Life and Writings of Benjamin Franklin, Written by Himself. New York: Mahlon Day, 1839.

Other Objects

Artist unknown, after a concept by Franklin. *La Grande Bretagne mutilée / Das verstümelte Britanien,* n.d. Colored engraving and etching.

Visiting Cards and Invitations Sent to Franklin in London and Paris, n.d.

Note from Secretary to the French King to Franklin, n.d.

Invitation to Franklin for a French Masonic Meeting, 1780s. Printed sheet with manuscript additions.

Jean-Honoré Fragonard

Le Docteur Francklin couronné par la Liberté, n.d. Aquatint and etching.

E. Gasteklu

Bronze Case for Volume of Catherine II's Decrees (Nakaz), 1776. Bronze. RAS

Christian Gottfried Geisler

Herbarium Illustration: "Catherinea Sublimus" (Macropodium nivale Pallas), 1793–94. Watercolor, graphite, and ink on paper. RAS

Herbarium Illustration: Didier's Tulip (Tulipa gesneriana), n.d. Watercolor and graphite on paper. RAS

Herbarium Specimen and Illustration: "Cheiranthi pinnatifides," n.d. Dried plant specimens mounted on paper, with watercolor, graphite, and ink illustrations. RAS

Aleksandr Bogdanov Gildebrand

Kovsh (Ladle), 1793. Silver gilt. Hillwood

Jean-Baptiste Greuze

Portrait of Benjamin Franklin, 1777. Oil on canvas.

Homann Erben (Homann Heirs)

America Septentrionalis a Domino d'Anville in Gallis edita nunc in Anglia (Map of North America). Nuremberg, 1777. Colored engraving. FLP

Ozias Humphrey

Miniature Portrait of Ekaterina R. Dashkova, ca. 1770. Watercolor and gouache on ivory, in gold-cased carnet by Jean-Jacques Prevost, 1764–65. Hermitage

Imperial Porcelain Factory, St. Petersburg

Bust of Catherine II, ca. 1810. Biscuit porcelain. Hillwood

Monteith (Bowl for Chilling Wine Glasses) from the Dowry Service of Grand Duchess Maria Pavlovna, 1799–1802. Hard-paste porcelain. Hillwood

Monteith (Bowl for Chilling Wine Glasses) with Views of Italy, from the Cabinet Service, 1796–99. Hard-paste porcelain. Hillwood

Imperial Academy of Sciences and Arts in St. Petersburg

Membership Diploma for Benjamin Franklin, November 2, 1789. Engraving with handwriting. BPL

Nova Acta, vol. 7. St. Petersburg: Imperial Academy, 1793.

Engraving Workshop. *A Display of the Fireworks and Illuminations for the New Year's Celebration, 1754.* Engraving. Kunstkamera

Imperial Academy of the Russian Language
Dictionary of the Academy of the Russian Language (Slovar' Akademii Rossiiskoi), vol. 6. St. Petersburg: Imperial Academy of Sciences, 1794. RAS

Thomas Jefferson
Manuscript Draft of the Declaration of Independence, 1776.

Andrei Kazachinsky
After Jean Louis de Veilly and Mikhail Makhaev. *Catherine II Receiving the Emissaries of Sultan Mustafa III,* 1796, printed 1857 (after 1794 drawing). Chisel etching. NYPL

Dmitrii G. Levitskii
Portrait of Ekaterina R. Dashkova, 1784. Oil on canvas. Hillwood

J. C. LeVasseur
After Antoine Borel. *L'Amérique Indépendante,* 1778. Engraving and etching.

Meriwether Lewis and William Clark
Checker Lily (Fritillaria affinis or *Fritillaria lanceolata),* collected 1806. Plant specimen. APS at ANSP

Antelope Bitterbrush (Purshia tridentata), collected 1806. Plant specimen. APS at ANSP

Canada Anemone (Anemone canadensis), collected 1804. Plant specimen. APS at ANSP

Lomonosov Color Glass Factory
Seven Glass Fragments, eighteenth century. Colored glass. Kunstkamera

Matthäus Albrecht (Matthew Albert) Lotter
A Plan of the City and Environs of Philadelphia. [Augsburg?], 1777. Colored engraving.

Tobias Conrad Lotter
Topographical Description of Petropolis, the Seat of the Muscovite Empire, in the Year 1744 (Topographia sedis Imperatoriae Muscovitarum Petropolis anno 1744). [Augsburg?], 1744. Colored engraving. LC

Jean-Hyacinthe Magellan

Facsimile of *Letter to Imperial Russian Academy of Sciences,* Paris, April 4, 1783. RAS

Mikhail Makhaev

Map of St. Petersburg (from *Plan stolichnago goroda Sanktpeterburga s izobrazheniem znatneishikh onago prospektov),* 1753. Engraving. NYPL

View down the Neva between the Winter Palace and the Academy of Sciences, 1753. Engraving. NYPL

Alexei N. Maximow

After unknown artist. *Portrait of Ekaterina R. Dashkova as Director of the Imperial Academy of Sciences and Arts,* 2003. Oil on canvas.

Johann Conrad Mayr

After Johann Georg Mayr. *View of Kir'ianovo, Ekaterina R. Dashkova's Dacha in the St. Petersburg Suburbs, with Serfs,* late eighteenth century. Colored aquatint. GIM

Domenico Menconi

Portrait Bust of Benjamin Franklin, ca. 1853. Marble.

D. T. Mussard

Pendant Watch with Cipher of Catherine II, 1786–96. Gold, diamonds, and rubies (Mussard, watchmaker; jeweler unknown). Hillwood

Jean-Baptiste Nini

Portrait Medallion of Benjamin Franklin in a Fur Cap, 1777. Terracotta. FI

Peter Simon Pallas

Flora Rossica (The Flora of Russia), vol. 1. St. Petersburg: J. J. Weitbrecht, 1784. PHS

Charles Willson Peale

After David Martin. *Portrait of Benjamin Franklin,* 1772 (after 1767 original). Oil on canvas.

Leopold Pfisterer

Star (Zvezda) of the Order of St. Catherine, 1792–97. Gold, enamel, and diamonds (attributed to Pfisterer). Hillwood

Potiomkin [Potemkin] Glassworks
Wineglass with Cyrillic Initials GMP, 1777–96. Green glass with silvering. Hillwood

Jean-Baptiste Le Prince
Experiment with Natural Electricity, ca. 1765. Pen, ink, and brown wash on paper. Rosenbach

Interior of a Russian Dwelling During the Night, ca. 1764. Pen, ink, pink and brown wash on paper. Rosenbach

Punishment by Flogging, 1766. Pen, ink, and gray wash on paper. Rosenbach

Punishment by Knouting (1ˢᵗ Degree), 1766. Pen, ink, and gray wash on paper. Rosenbach

Punishment by Knouting (2ⁿᵈ Degree), 1766. Pen, ink, and gray wash on paper. Rosenbach

Russian Dance, 1754. Pen, ink, and gray wash on paper. Rosenbach

Russian Supper, 1764. Pen, ink, pink and brown wash on paper. Rosenbach

Frederick Pursh
Illustration of Lewis & Clark Plant Specimen: Canada Anemone (Anemone canadensis), after 1806. Watercolor and graphite on paper.

Illustration of Lewis & Clark Plant Specimen: Checker Lily (Fritillaria affinis or *Fritillaria lanceolata),* after 1806. Watercolor and graphite on paper.

Illustration of Lewis & Clark Plant Specimen: Antelope Bitterbrush (Purshia tridentata), after 1806. Watercolor and graphite on paper.

Giacomo Quarenghi
Facsimile of *The Stage of the Hermitage Theater, St. Petersburg,* 1787. Engraving. GIM

Aleksandr N. Radishchev
Journey from St. Petersburg to Moscow (Puteshestvie iz Peterburga v Moskvu). St. Petersburg, 1790. LC

Christoph Melchior Roth
New Map of the Capital City and Fortress of St. Petersburg (Novoi plan stolichnago goroda i krieposti Sanktpeterburga), 1776. Colored engraving. LC

Royal Porcelain Factory, Berlin
Coffee Set with Views of Berlin, 1790s. Porcelain with overglaze polychrome painting, paste, and gilding. GIM

Royal Porcelain Factory, Sèvres, France
Nicholas-Pierre Pithou the Younger, and Henri-Martin Prévost, decorators. *Cup and Saucer with Portrait of Benjamin Franklin,* 1778. Hard-paste porcelain with enamel and gilt decoration. PMA

Christian Schuessele
Franklin before the Lords in Council, 1774. New York: Thomas Kelly, n.d. Colored engraving.

Nicholas Scull and George Heap
An East Prospect of the City of Philadelphia, 1754 (1ˢᵗ ed.). Engraving. HSP

Map of Philadelphia and Parts Adjacent, 1777. Engraving.

R. W. Seale
A Map of North America with the European Settlements, ca. 1744. Engraving.

Georgiana Shipley
Letter to Benjamin Franklin. Twyford, January 6, 1781.

Gavriil I. Skorodumov
Portrait of Ekaterina R. Dashkova, 1777. Sepia stipple engraving. GIM

Portrait of Ekaterina R. Dashkova with Her Son Pavel and Daughter Anastasiia in England, late 19th century (reproduction of 1777 original). Engraving. GIM

Charles Stanhope
Principles of Electricity: Containing Divers New Theorems and Experiments. London: P. Elmsly, 1779.

Ezra Stiles
Meteorological Memoir Sent to Mikhail V. Lomonosov, Newport, February 14, 1765.

François Marie Suzanne
Portrait Statuette of Benjamin Franklin, 1783. Terracotta. Walters

P. C. Varlé
To the Citizens of Philadelphia, This Plan of the City and its Environs is Respectfully Dedicated, 1802. Colored engraving.

A. Vil'brekht
General Map of the Russian Empire Divided into Forty-One Province (*Rossiiskoi Imperii na sorok odnu guberniiu razdel'onnoi),* 1790s. Colore

Josiah Wedgwood

Anti-Slavery Medallion, ca. 1788. Black and white jasper dip.

Francis Wheatley

Facsimile of *The Dublin Volunteers on College Green, 4th November 1779,* 1779–80. Oil on canvas. NGI

Robert Whitechurch

After Christian Schuessele. *Franklin at the Court of Saint James, 1774,* 1868. Engraving and etching.

Jonathan Williams

Thermometrical Navigation: Being a Series of Experiments. Philadelphia: R. Aitken, 1799.

Wistarburgh Glass House, Salem, N. J.

Bottle, eighteenth century. Green glass. SCHS

Bull's-Eye Pane, eighteenth century. Green glass. SCHS

Case Bottle, eighteenth century. Green glass. SCHS

Covered Bowl, eighteenth century. Green glass. SCHS

Glass Chunk, eighteenth century. Green glass. SCHS

Glass Tube for Electrical Experiments, eighteenth century. Green glass.

Glass Tube for Electrical Experiments, eighteenth century. Green glass. LCP

Slag (Leftover Glass), eighteenth century. Green glass. SCHS

Selected Dashkova Bibliography

The following is not a comprehensive bibliography on either Dashkova or Franklin. For Dashkova, it consists of published books and articles selected by the catalogue authors and the editor, focusing especially on English-language sources and on themes relating to the essays in this book. Other Dashkova sources, including most Russian-language publications and some manuscript materials, can be found in the authors' endnotes. For a thorough Dashkova bibliography, see *Mon histoire,* the French edition of Dashkova's memoirs (listed below). For an introduction to the vast literature on Franklin, see the 2005 exhibition catalogue *Benjamin Franklin: In Search of a Better World,* edited by Page Talbott (also listed below). A few Franklin sources are also cited in the authors' endnotes.

Amoia, Alba, and Bettina L. Knapp, eds. *Great Women Travel Writers: From 1750 to the Present.* New York: Continuum, 2005.

Bisha, Robin, Jehanne M. Geith, Christine Holden, and William G. Wagner, eds. *Russian Women, 1698–1917: Experience and Expression, An Anthology of Sources.* Bloomington and Indianapolis: Indiana University Press, 2002.

Bradford, Martha Wilmot. *The Russian Journals of Martha and Catherine Wilmot . . . 1803–1808.* Edited by the Marchioness of Londonderry and H. M. Hyde. London: Macmillan, 1935.

Catherine II, Empress of Russia. *Memoirs of Catherine the Great.* Translated by Katharine Anthony. New York: Alfred A. Knopf, 1927.

Cross, Anthony G. *"By the Banks of the Thames": Russians in Eighteenth Century Britain.* Newtonville, Mass.: Oriental Research Partners, 1980.

———. "Contemporary British Responses (1762–1810) to the Personality and Career of Princess Ekaterina Romanovna Dashkova." *Oxford Slavonic Papers,* n.s., 27 (1994): 41–61.

Cross, Anthony G., and G. S. Smith. *Eighteenth-Century Russian Literature, Culture, and Thought:*

A Bibliography of English-Language Scholarship and Translations. Newtonville, Mass.: Oriental Research Partners, 1984.

Crowe, Nicholas. "Ekaterina Romanova Dashkova, 1743–1810." In *Reference Guide to Russian Literature,* edited by Neil Cornwell, 238–40. London: Fitzroy Dearborn Publishers, 1998.

Dashkova, Ekaterina Romanovna. *The Memoirs of Princess Dashkova: Russia in the Time of Catherine the Great.* Translated and edited by Kyril Fitzlyon. Durham: Duke University Press, 1995.

———. *Memoirs of the Princess Daschkaw, Lady of Honour to Catherine II, Empress of All the Russias* . . . . Edited by Mrs. W. Bradford [Martha Wilmot Bradford]. 2 vols. London: Henry Colburn, 1840.

——— [Princesse Dachkova]. *Mon histoire: Mémoires d'une femme de lettres russe à l'époque des Lumières.* Edited by Alexander Woronzoff-Dashkoff, Catherine Le Gouis, and Catherine Woronzoff-Dashkoff. Paris: L'Harmattan, 1999.

De Madariaga, Isabel. *Catherine the Great: A Short History.* New Haven: Yale University Press, 1990.

———. *Russia in the Age of Catherine the Great.* New Haven: Yale University Press, 1981.

Diderot, Denis. *Correspondance.* Vol. XI, avril 1771–déc. 1771. Edited by Georges Roth. Paris: Éditions de minuit, 1964.

———. "Sur la princesse Daschkoff." In *Oeuvres complètes de Diderot,* 17: 487–90. Paris: Garnier frères, 1876.

Dvoichenko-Markoff, Eufrosina. "The American Philosophical Society and Early Russian-American Relations." *Proceedings of the American Philosophical Society* 94 (1950): 549–610.

———. "Benjamin Franklin, the American Philosophical Society, and the Russian Academy of Science." *Proceedings of the American Philosophical Society* 91 (1947): 250–57.

Echeverria, Durand. *Mirage in the West: A History of the French Image of American Society to 1815.* Princeton, N.J.: Princeton University Press, 1957.

Echeverria, Durand, and Everett C. Wilkie, Jr. *The French Image of America: A Chronological and Subject Bibliography of French Books Printed Before 1816 Relating to the British North American Colonies and the United States.* 2 vols. Metuchen, N.J.: Scarecrow Press, 1994.

Gay, Peter. *The Enlightenment: An Interpretation.* Vol. 2, *The Science of Freedom.* New York: Alfred A. Knopf, 1969.

Gertsen, Aleksandr [Alexander Herzen]. "Princess Yekaterina Romanova Dashkov." *The Pole Star,* 1857. Reprinted in *My Past and Thoughts: The Memoirs of Alexander Herzen,* 4: 1585–1646. Translated by Constance Garnett. Revised by Humphrey Higgens. New York: Alfred A. Knopf, 1968.

Giroud, Vincent. *St. Petersburg: Portrait of a Great City.* New Haven, Conn.: Beinecke Rare Book and Manuscript Library, Yale University, 2003.

Gordin, Michael D. "The Importation of Being Earnest: The Early St. Petersburg Academy of Sciences." *Isis* 91 (2000): 1–31.

Haumant, E. *La culture française en Russie.* Paris: Hachette, 1913.

Heldt, Barbara. *Terrible Perfection: Women and Russian Literature.* Bloomington: Indiana University Press, 1987.

Hindle, Brooke. *The Pursuit of Science in Revolutionary America, 1735–1789.* Chapel Hill: University of North Carolina Press, 1956.

Hyde, H. Montgomery. *The Empress Catherine and Princess Dashkov.* London: Chapman and Hall, 1935.

Igumnova, T. G., ed. *Sei put' tebe prinadlezhit: Kniagine Ekaterine Romanovne Dashkovoi posviaschaetsa.* Moscow: Gos. istoricheskii muzei, 2003.

Kolchin, Peter. *Unfree Labor: American Slavery and Russian Serfdom.* Cambridge, Mass.: Harvard University Press, Belknap Press, 1987.

Larivière, Ch. de. *Catherine II et la révolution française.* Paris: H. Le Soudier, 1895.

Longmire, Robert Argent. "Princess Dashkova and the Intellectual Life of Eighteenth-Century Russia." Master's thesis, University of London, 1955.

Mailloux, Luc. "La princesse Daschkoff et la France." *Revue d'histoire diplomatique* 95 (1981): 5–25.

Marrese, Michelle Lamarche. *A Woman's Kingdom: Noblewomen and the Control of Property in Russia, 1700–1861*. Ithaca, N.Y.: Cornell University Press, 2002.

Masson, Charles François Philibert. *Mémoires secrets sur la Russie, et particulièrement sur la fin du règne de Catherine II et le commencement de celui de Paul I . . . .* Vol. 2. Paris: Charles Pougens, 1800.

Meehan-Waters, Brenda. "Catherine the Great and the Problem of Female Rule." *Russian Review* 34 (1975): 293–307.

Menshutkin, Boris N. *Russia's Lomonosov: Chemist, Courtier, Physicist, Poet*. Princeton, N.J.: Princeton University Press, 1952.

Nash, Carol S. "Educating New Mothers: Women and the Enlightenment in Russia." *History of Education Quarterly* 21 (1981): 301–16.

Neverov, Oleg. "Gems from the Collection of Princess Dashkov." *Journal of the History of Collections* 2, no. 1 (1990): 63–68.

Odom, Anne, and Liana Paredes Arend. *A Taste for Splendor: Russian Imperial and European Treasures from the Hillwood Museum*. Alexandria, Va.: Art Services International, 1998.

Parkinson, John. *A Tour of Russia, Siberia and the Crimea, 1792–1794*. Edited by William Collier. London: Franc Cass, 1971.

Pavlenko, Nikolai. "The Other Catherine." *Russian Life* 42 (December–January 1999): 51–53.

Petrova, Yevgenia, ed. *St. Petersburg: A Portrait of the City and its Citizens*. St. Petersburg: State Russian Museum / Palace Editions, 2003.

Radishchev, Aleksandr Nikolaevich. *A Journey from St. Petersburg to Moscow*. Translated by Leo Wiener. Edited by Roderick Page Thaler. Cambridge, Mass.: Harvard University Press, 1958.

Raney, Sherri Thompson. "A Worthy Friend of Tomiris: The Life of Princess Ekaterina Romanova Dashkova." Ph.D. diss., Oklahoma State University, 1993.

Rice, Tamara Talbot. *A Concise History of Russian Art*. New York: Praeger, 1963.

Rogger, Hans. *National Consciousness in Eighteenth-Century Russia*. Cambridge, Mass.: Harvard University Press, 1960.

Roosevelt, Priscilla R. *Life on the Russian Country Estate: A Social and Cultural History*. New Haven, Conn.: Yale University Press, 1995.

"The Russian Academy." *The Edinburgh Magazine* 47 (April 1785): 304–7.

Scarisbrick, Diana. "Companion to a Russian Princess: Martha Wilmot's Green Book." *Country Life* 169 (January 8, 1981): 76–78.

Schlegelberger, Günther. *Die Fürstin Daschkowa: Eine biographische Studie zur Geschichte Katharinas II*. Berlin: Junker und Dünnhaupt Verlag, 1935.

Talbott, Page, ed. *Benjamin Franklin: In Search of a Better World*. New Haven, Conn.: Yale University Press, 2005.

Tishkin, G. A. "A Female Educationalist in the Age of Enlightenment: Princess Dashkova and the University of St. Petersburg." *History of Universities* 13 (1984): 137–52.

Venturi, Franco. *The End of the Old Regime in Europe, 1768–1776: The First Crisis*. Translated by R. Burr Litchfield. Princeton, N.J.: Princeton University Press, 1989.

———. *The End of the Old Regime in Europe, 1776–1789: The Great States of the West*. Translated by R. Burr Litchfield. Princeton, N.J.: Princeton University Press, 1991.

Vowles, Judith. "The 'Feminization' of Russian Literature: Women, Language, and Literature in Eighteenth-Century Russia." In *Women Writers in Russian Literature*, edited by Toby W. Clyman and Diana Greene, 35–60. Westport, Conn.: Greenwood Press, 1994.

Walpole, Horace. *The Yale Edition of Horace Walpole's Correspondence*. Edited by W. S. Lewis et al. 48 vols. New Haven, Conn.: Yale University Press, 1937–83.

Whittaker, Cynthia Hyla, Edward Kasinec, and Robert H. Davis, Jr., eds. *Russia Engages the World, 1453–1825*. Cambridge, Mass.: Harvard University Press; New York: New York Public Library, 2003.

Wolff, Larry. *Inventing Eastern Europe: The Map of Civilization on the Mind of the Enlightenment*. Stanford, Calif.: Stanford University Press, 1994.

Woronzoff-Dashkoff, Alexander. "Disguise and Gender in Princess Dashkova's *Memoirs*." *Canadian Slavonic Papers* 33, no. 1 (1991): 62–74.

———. "Ekaterina Romanova Dashkova." In *Dictionary of Russian Women Writers*, edited by Marina Ledkovsky, Charlotte Rosenthal, and Mary Zirin, 142–44. Westport, Conn.: Greenwood Press, 1994.

———. "Ekaterina Romanovna Dashkova." In *Early Modern Russian Writers, Late Seventeenth and Eighteenth Centuries*, edited by Marcus C. Levitt, 65–69. Detroit: Gale Research, 1995.

———. "Ekaterina Romanovna Dashkova." In *Russian Women Writers*, edited by Christine Tomei, 1: 29–42. New York: Garland Publishers, 1999.

———. "Princess E. R. Dashkova: First Woman Member of the American Philosophical Society." *Proceedings of the American Philosophical Society* 140, no. 3 (1996): 406–17.

———. "Princess E. R. Dashkova's Moscow Library." *Slavonic and East European Review* 72 (January 1994): 60–71.

——— et al., eds. *Ekaterina Romanovna Dashkova: issledovaniia i materialy*. St. Petersburg: Dmitrii Bilanin, 1996.

Wortman, Richard. "Property Rights, Populism, and Russian Political Culture." In *Civil Rights in Imperial Russia*, edited by Olga Crisp and Linda Edmondson, 13–32. Oxford: Clarendon Press, 1989.

Index